*The Consequences
of the Peace*

The Twentieth Century Fund is a research foundation undertaking timely analyses of economic, political, and social issues. Not-for-profit and nonpartisan, the Fund was founded in 1919 and endowed by Edward A. Filene.

The Consequences
of the Peace

The New Internationalism and
American Foreign Policy

James Chace

OXFORD UNIVERSITY PRESS
New York   Oxford

Oxford University Press

Oxford   New York   Toronto
Delhi   Bombay   Calcutta   Madras   Karachi
Kuala Lumpur   Singapore   Hong Kong   Tokyo
Nairobi   Dar es Salaam   Cape Town
Melbourne   Auckland   Madrid

And associated companies in
Berlin   Ibadan

Copyright © 1992 by The Twentieth Century Fund

First published in 1992 by Oxford University Press, Inc.,
200 Madison Avenue, New York, New York 10016

First issued as an Oxford University Press paperback, 1993

Oxford is a registered trademark of Oxford University Press

Library of Congress Cataloging-in-Publication Data
Chace, James.
The consequences of the peace: the new internationalism
and American foreign policy / James Chace
     p.   cm.   "A Twentieth Century Fund book."
Includes bibliographical references and index.
ISBN 0-19-507411-4
ISBN 0-19-508354-7 (PBK)
  1. United States—Foreign relations—1989.   I. Title.
E881.C42   1992   327.73—dc20   91-35997

10 9 8 7 6 5 4 3 2 1

Printed in the United States of America

For Sidney

Foreword

The past few years have been marked by the brutal repression of a remarkable outburst of free thinking in China, the liberation of the East European satellites of the Soviet bloc, the reunification of Germany, the apparent collapse of the central Soviet state, the outlawing of communism in the Soviet Union, an American-led coalition of nations punishing Iraq for aggression, the failure of a hard-line coup in the Soviet Union, and the start of the regional peace talks between the Israelis and the Arabs.

Small wonder that much of the analysis and writing currently available on American foreign policy is tentative and uncertain. At the same time, policymakers have a sharpened appetite for bold and far-reaching responses to the current situation.

For more than four decades the ideas of liberal internationalism have guided the foreign policy of the United States. Given recent events, Americans are bound to ask whether this doctrine is adequate for the new world we now inhabit. The domestic problems of this nation will make the question all the more pressing. Why should the United States remain engaged abroad when public investment is badly needed at home?

In the pages that follow, James Chace, former managing

editor of *Foreign Affairs* and noted teacher and writer on foreign policy issues, confronts this challenge to our traditional foreign policy and advocates a "new internationalism" that serves American interests and gives hope of a world order that is both peaceful and prosperous. Chace believes that the coming global era will not be dominated by any single nation. Acting as first among equals, the United States should lead the major powers toward reforming and revitalizing the international economic institutions set up after the last world war.

The future will also require, in Chace's judgment, American attention to complicated balances of power throughout the developed world. In this way, Chace poses a challenge to American leaders who, it seems, are often more comfortable responding to crises than actively directing and shaping change.

Over the past seventy years, the Twentieth Century Fund has sponsored many studies of American foreign policy, including, most recently, Richard Ullman's *Securing Europe,* an examination of the problem of European security in the new world, and Bennett Kovrig's *Of Walls and Bridges,* an examination of the effects of American foreign policy decisions on Europe. The Fund has a number of ongoing projects in this area as well, but I suspect that few will prove more timely than this book by James Chace.

Perhaps, when the present era is recalled, some future historian will ask: "How did those who shaped foreign policy find time to keep abreast of events, let alone think through new approaches to the rapidly changing world?" James Chace's work is likely to be among the examples used to answer that question. We are grateful to him for undertaking the effort and pressing forward in the face of an ever shifting world scene.

Richard C. Leone, President
The Twentieth Century Fund
November 1991

Acknowledgments

This book could not have been written without the aid and encouragement of the Twentieth Century Fund. I have benefited enormously from the editorial suggestions of Brewster Denny, Beverly Goldberg, Roger Kimball, and Richard C. Leone, president of the Fund.

I have also found invaluable the detailed editorial comments of Nicholas X. Rizopoulos, Charles Kupchan, and Tony Smith.

I have benefited enormously from the editorial skills of Linda Wrigley and from the research assistance of Kate Doyle.

Valerie Aubry and David Roll of Oxford University Press safely guided the book to publication.

Some passages in the present work originally appeared in the following publications: *Foreign Affairs, Foreign Policy, The New York Review of Books, The New York Times Magazine,* and in a chapter of the book *Theory in America, America in Theory.*

I have dedicated this book to Sidney Blumenthal, who has been unfailingly generous with his time and who wisely urged me never to overlook the essential connections between domestic politics and American foreign policy.

Contents

The events of the coming year will not be shaped by the deliberate acts of statesmen, but by the hidden currents, flowing continually beneath the surface of political history, of which no one can predict the outcome. In one way only can we influence these hidden currents—by setting in motion those forces of instruction and imagination which change *opinion*. The assertion of truth, the unveiling of illusions, the dissipation of hate, the enlargement and instruction of men's hearts and minds, must be the means.

John Maynard Keynes
The Economic Consequences of the Peace

The Consequences
of the Peace

Introduction:
The New
Internationalism

THE Hungarians who lived in Transylvania, that province of the Austro-Hungarian empire that had been ceded to Rumania after World War I, had long chafed under Bucharest's efforts to curb any signs of Hungarian nationalism. In the 1980s, however, attacks on Rumanian repression by one pastor, Laszlo Tokes, in the city of Timisoara, especially aroused the ire of Rumania's communist regime. Finally, in the fall of 1989, a Rumanian court ordered the eviction of the Reverend Mr. Tokes on December 15. That morning, many citizens of Timisoara—and not all of them ethnic Hungarians—gathered in front of the church to oppose Tokes' eviction. For the next two days there was rioting in the city to protest the excesses of the regime that had been headed by the corrupt and tyrannical Nicolae Ceausescu for twenty-four years. Then on Sunday, December 17, armed men in civilian clothes, members of Ceausescu's dreaded security apparatus, the Securitate, opened fire on the crowd. For three days the confrontation continued, and almost 700 people died.

But at the end of that period, the rebellious populace gained control of one of the city's largest factories, and forced the Securitate to release prisoners held in the city jail. More important, news of the Timisoara uprising reached Bucharest. Ceausescu decided to address the nation on television, at which time he insisted that the uprising, "organized and unleashed in close connection with reactionary, imperialist, irredentist, chauvinist circles," be put down. He called for a rally of support the next day.

At the demonstration that had been orchestrated to show that the people were loyal to the regime, the unthinkable happened. Cries of "Timisoara, Timisoara" echoed throughout the crowd, and a banner was unfurled carrying the message, "Down with Ceausescu." The aged despot was stunned by what was happening. He stopped speaking from his balcony; when he resumed, the shrieks of the crowd, the shouted slogans of "Yesterday Timisoara, today Bucharest" and "The Army is with us" signaled the death knell of the Ceausescu regime. The people knew what had been happening elsewhere in Eastern Europe: the Berlin Wall was being torn down, and in Hungary, Poland, and Czechoslovakia the peoples had turned out their communist rulers.[1]

Of all those transforming events that took place in the momentous year of 1989, the downfall and execution of Nicolae Ceausescu and his wife on Christmas Day were perhaps the most significant. Ceausescu was the only East European communist leader who had used bullets to try to destroy the popular forces arrayed against him. He failed, but it was a near thing. In those bloody days before Christmas, the ruling group, with the army's backing, had emerged to take over the direction of the revolution, and were at one point prepared to ask for Soviet military assistance to prevent a counterattack by forces still loyal to Ceausescu. Had they done so and had the Soviet military responded, it would have been with Washington's blessing.

For while the battle for Rumania was still hanging in the balance, the American Secretary of State James Baker endorsed the idea of Soviet military intervention. On Christmas Eve, Baker said that the United States would approve the use of Soviet or Warsaw Pact forces against the repressive Ceausescu dictatorship. "If the Warsaw Pact felt it necessary to intervene on behalf of the opposition," he declared, the United States "would support that action."[2] With these words, he reversed the

U.S. policy of the containment of the Soviet Union that had lasted more than four decades. The Cold War, which had begun over Soviet control of Eastern Europe, was truly over.

Halfway across the globe, that same Christmas week, 25,000 American troops invaded Panama. The ostensible *casus belli* was the killing of an American soldier and the roughing up of American personnel by thugs of the Panamanian strongman, General Manuel Antonio Noriega. The deeper reasons for the intervention lay in Washington's traditional hegemonic response when challenged in the Caribbean-Central America region. Despite the end of the Cold War and the growing willingness of the Soviet Union to relinquish any significant role in the Western Hemisphere, the United States showed itself only too ready to project force to further the national interest.

For years the United States had supported the existence of the Panamanian Defense Force (PDF). As Washington saw it, the PDF would prove a force for stability, which had long been the overriding goal of American policy in the hemisphere. Noriega, who had made his career in the intelligence services of the PDF, was also contracted to the Central Intelligence Agency. But in 1987, when one of his once-trusted officers accused Noriega of dealing with the Colombian drug cartels and of personal corruption, the Reagan administration decided that Noriega was a political embarassment and of much less use to the United States.

Noriega proved a wily adversary. Although the United States suspended all economic and military aid to Panama in the summer of 1987 and applied further, though ineffective, economic sanctions in the spring of 1988 after Noriega was indicted by a Miami grand jury for drug trafficking, the general easily held on. In October 1989, a coup mounted against him by dissident officers failed. This time, there would be no failure. On December 20, massive air and ground attacks—Operation Just Cause—ensured that Noriega's regime would fall. But the significance of the intervention went far beyond the toppling of a dictator. It was the first time in forty years that the United States had used force to intervene without justifying its use by referring to the Cold War.

From Korea to Vietnam to tiny Grenada, Washington had seen its interventions as dictated by the exigencies of the global

containment of the Soviet Union. The world was now turned
upside down. We were now prepared to cooperate with the
Soviets to impose order outside well-established borders. We
also intended to intervene unilaterally to protect our interests
as we saw them, especially in our own sphere of influence in the
Western Hemisphere, a region the Soviets were now prepared
to leave alone. Above all, we were ready to demonstrate to the
world that the United States was self-confident enough to use
its military forces far from its shores, while the Soviet Union
expended its energies trying to prevent its own dissolution.

Less than a year after the Panamanian intervention, the
United States sent almost half a million troops to Saudi Arabia
and the Persian Gulf to lead an attack on Iraq's Saddam Hussein, who was trying to establish Iraq as the dominant power in
the Middle East. Yet at the very moment when hostilities against
Iraq began, in January 1991, the Soviets were using tanks in
Lithuania to prevent the tiny Baltic republic from formally
breaking away from the Soviet Union. Despite the cooperative
relationship among the great powers and, in particular, despite
the amity that existed between Washington and Moscow, the
exercise of military force appeared undiminished.

Nonetheless, thanks to the end of the Cold War, the society
of nations found itself nearing the condition that the drafters
of the United Nations charter had hoped for at the end of
World War II. For forty-five years, the rivalry between the
United States and the Soviet Union had prevented the U.N.
Security Council from carrying out the provisions on collective
security that the Security Council was orginally designed to per-
form.[3] In the crisis in the Persian Gulf, provoked by Iraq's invasion of Kuwait in August 1990, the Security Council voted economic sanctions to force Iraq to withdraw from Kuwait.
Although the rhetoric used to justify the Council's action
stressed the need to punish aggression, the real reason that the
great powers were able work together in this instance was the
time-honored need to reestablish the status quo in the Gulf in
order to ensure reasonable access to oil.

The end of the Cold War, followed by the victory of the
U.S.-led military coalition in the Gulf war, was a defining
moment for U.S. foreign policy. It was clear that the old internationalism that had characterized U.S. policy during the Cold
War was no longer applicable. That internationalism had

grown out of the felt need to combat the twin ideologies of fascism and communism; moreover, it had arrived on the world's stage just as America switched from leading an antifascist alliance to leading an anticommunist one, and so made it possible for former conservatives to take part in an internationalist consensus.[4] It believed that the American mission was to organize the world along lines that would mirror American values and aspirations. An open trading system in which the United States provided the world with the money needed to lubricate the global economy would bring prosperity to those nations that chose to participate. An alliance system in Europe and the Pacific would serve as the twin anchors of a strategy designed to contain an expanding—and ideologically threatening—Soviet Union.

With the collapse of the communist system in Eastern Europe and more important, the liberation of the Soviet Union from communism after the failed coup of August 1991, the world seemed blessedly free from ideological struggle. Democracy and freedom of economic choice are aspirations that are now generally shared by Russia and America and by most of the Third World. The United States thus finds itself deprived of its national mission and its self-justification for over four decades.[5] The questions before us now are "What should the United States do?" and, of equal importance, "What should the United States be?"

Can there be a new internationalism that will sustain America's foreign and domestic aspirations? If so, it will have to be based on a new American perception of our economic, military, and geopolitical position at the end of the twentieth century. Above all, it must take account of a new reality—the disappearance of superpower status in the world.

Since the end of the Cold War, there have been a number of points of view advanced, none of which actually define the new internationalism, although they often appear in the guise of internationalist rhetoric. There are roughly three schools of thought that pretend to provide new answers to the changed conditions of international life that affect U.S. perceptions of its military and economic security. One school might be labeled "neo-isolationist," though it sometimes prefers to call itself "interest-based" foreign policy.[6] A second approach, which might best be termed "triumphalism," believes that America's leading role in the Gulf war has left it as the only superpower. Those who cling to this vision of America-as-number-one would put the United States in the role of global enforcer, committed to an unending policy of unilateral interventionism. The third approach, which President Bush has called for, is a New World Order, a "world in which nations recognize shared responsibility for freedom and justice" and "where the rule of law supplants the rule of the jungle."[7] Beneath this neo-Wilsonian rhetoric, however, Bush's call is for the status quo, a latter-day version of the Pax Americana we have known for nearly half a century.

At first blush, the neo-isolationist stance can seem appealing. Its central premise is that the United States can no longer afford an internationalist foreign policy based on military and economic prowess. A defense budget for the 1990s that still approaches $300 billion a year is no longer sustainable for a nation that supports an ever-greater burden of entitlements; that suffers from a decaying infrastructure, a faltering educational system, a low savings rate, and an unwillingness to invest for the long term. By running large, chronic federal deficits for over a decade, the United States has had to borrow in the international capital markets. This pushed up the cost of capital and led to a fall in U.S. exports, while giving Americans a new taste for relatively inexpensive imports. In this respect, the United States has proved profligate with its resources and can hardly be expected to pursue a policy of global engagement, as it did in the heady years of the Cold War.

While it would be difficult to fault an analysis that called on the American people to reorder their national priorities, the neo-isolationist policy goes further and draws faulty conclusions that ignore U.S. dependence on a global economy. As Alan Tonelson spelled it out in a July 1991 article in *The Atlantic* attacking internationalism, the United States should "rule out economic initiatives deemed necessary for the international system's health if those initiatives wound up siphoning more wealth out of this country . . . than they brought in."[8] That sounds fine, except for the fact that a healthy international economic system is indispensable for American prosperity.

America's role is to devise new institutions to ensure a healthy functioning of this system rather than, as Tonelson would have it, to secure "the maximum degree of freedom of action and self-reliance in a still dangerous world." Our markets, after all, are to be found in East Asia, the European Community, and Latin America, where most of our trade goes. In this respect, America's security is not tied up solely with "such trouble spots [that bear] directly on America's core security (for example, Mexico and the Caribbean Basin)." In short, a new, noninternationalist approach would "decouple America's security from that of its allies" and "keep unfriendly foreign powers out of the hemisphere by using and threatening to use force unilaterally."[9]

Nothing would be more self-defeating for a global economic power than for the United States to lodge itself in the cocoon of the Western Hemisphere. In this posture, the United States would presumably do little to maintain the balance of power in East Asia, or to provide the reassurance that the Europeans seek that neither Germany nor Russia will ever again threaten the peace of the West. Should the relative, if uneasy, equilibrium that now obtains in East Asia and the Western Pacific be disturbed by an aggressive Japan, a resurgent China, or a reactionary Russia, the isolationist argument implies that the United States would presumably rest safe and sound in its hemisphere, its trade unmolested, its economy and its populace protected by the deterrent value of its nuclear armory.

If neo-isolationism is shortsighted, triumphalism is quixotic—and dangerous. In the wake of the Gulf war, when the United States ran the show, with the often grumbling acquiescence of Russia and China, it might seem that the United States is indeed the only remaining superpower. According to the triumphalists, the old bipolar world has been succeeded by a new world that is not multipolar but unipolar. In an article that appeared in *Foreign Affairs* early in 1991, columnist Charles Krauthammer has called this decade "the unipolar moment." He dismisses the notion that Europe's (now including a unified Germany) and Japan's economic power would give them great-power status. Because of their unwillingness to send significant military forces to the Gulf war, the argument goes, Germany and Japan do not qualify as rivals of the United States. The use of the United Nations as a mechanism to organize an effective resistance to Iraqi aggression against Kuwait is similarly dismissed as "pseudo-multilateralism." Yet it is hard to imagine that the United States would have been able to dispatch a large expeditionary force to the Persian Gulf without Moscow's evident willingness to abandon its erstwhile ally Iraq when Saddam Hussein proved impervious to reason. As for the Germans and the Japanese, they virtually underwrote the cost of the war. If that does not confer great-power status on them, what does? Underwriting a war to protect one's vital interests is not necessarily the mark of weakness; it could also be construed as a sign of a clever prudence.

Looking to the future, Krauthammer believes that the United States need only show the will to support its new "unipolar status." Its main job: to destroy the so-called Weapons States, (i.e., small, Third World countries such as Libya, Iraq, and North Korea that might produce strategic nuclear weapons). He sets aside efforts to obtain a multilateral backing for such a policy, although the United States and the Soviet Union have long supported nonproliferation; moreover, should a North Korea, for example, brandish such weapons, a coalition of Russia and China would almost certainly form to compel North Korea to disarm. The "unipolarists"—like the old-fashioned unilateralists—evoke a world where the United States will be able to practice a "robust and difficult interventionism" and "unashamedly [lay] down the rules of world order and [be] prepared to enforce them."[10]

In addition, the unilateralists ignore the economic dimension of U.S. foreign policy. They fail to comprehend that it is far more important for the United States to cooperate with the other great powers to ensure a healthy functioning of the world economy than to expend its energies by unilaterally assuming the task of enforcing peace and stability throughout the world. There is a striking absence in their analysis of any concern with ends and means.

George Bush's appeal for a New World Order is neither unilateralist nor isolationist. It is rather a call for a universal alliance against disorder. The alliance would be led by the United States, though largely paid for by others. It claims not to be a latter-day version of a Pax Americana. "I would not call [the United States] the world's policeman, because there are certain areas where we wouldn't be in a position to act or want to act," the president said. "But we have a disproportionate responsibility for the freedom and security of various countries."[11] This sounds remarkably like the world order other Cold

War leaders of the United States have preferred. But, in one key aspect Bush differs from the early architects of postwar American foreign policy: he is less concerned with freedom and justice than with the status quo—as witness his policy to restore the Emir of Kuwait, not to recognize quickly the Baltic republics, or to challenge the Chinese on their occupation of Tibet. He has called for a New World Order, but his is a vision of a world that has not really changed. Even though he has called for the elimination of tactical nuclear weapons and the reduction of force, his notion of America's posture closely resembles our role as global gendarme.

The dangers that the Bush adminstration evokes are "instability" and "uncertainty."[12] The new world, no longer threatened by the great communist powers of China and Russia, is endangered by drug traffickers, Weapons States, and various forms of ethnic and religious turmoil. Indeed, in the absence of the superpower rivalry that characterized the Cold War, there may well be a more anarchic world, one in which one power will no longer be able to restrain its client for fear of retaliation by the other. Bush's belief that a New World Order will demonstrate that "brutality will go unrewarded and aggression will meet collective resistance"[13] rests on the proposition that the United States will garner its support for the rule of law from the members of the U.N. Security Council. Yet before, during, and after the Gulf war, the continent of Africa was wracked by several wars—in Ethiopia, Somalia, Liberia—toward which the Bush administration was relatively indifferent.

Moreover, much of the turmoil that threatens the status quo is likely to stem from internal struggles for ethnic, religious, and national identity within the Soviet Union or China (which, after all, incorporated Tibet against its will). Eastern Europe may continue to contend with ethnic struggles, as in Yugoslavia, that may well affect neighboring states. In the Middle East, religious groups vie for supremacy and identity within boundaries drawn by the former colonial powers, Britain and France. Yet for the past half-century it has been well nigh impossible to intervene in the name of order and stability in the internal affairs of other nations.

For Bush, the new adversary is instability, which demands a policeman rather than the leader of a coalition. "From principal custodian of freedom against a specific adversary," political scientists Robert W. Tucker and David C. Hendrickson have written, "America was now to become principal custodian of stability and order against any state threatening the tranquillity of the international system."[14] Bush did not see the United States win the Cold War in order to relinquish America's position as the enforcer of a Pax Americana.

The Gulf war, however, is not a portent of things to come. The reason the United Nations was able to play an important role in the crisis (and its aftermath), the reason the United States was able to send almost half a million troops to the region, let us not forget, was because of Soviet-American cooperation. Aggression was punished because the world community did not want to see a vital resource—oil—come under the sway of an aggressive and unpredictable tyrant. It is hard indeed to imagine the United States dispatching a massive expeditionary force elsewhere in the world to subdue a small but aggressive state and receive the support of most of the members of the United Nations and all of the members of the U.N. Security Council.[15]

The decades-long conflict with the Soviet Union provided the United States with the basis for a global foreign policy, so that, in the minds of American policymakers, as foreign policy analyst Michael Mandelbaum has written, "the various conflicts of the Cold War were all connected. The Greek civil war, the Korean War, the Vietnam War and others were seen as part of a global struggle against communism. Each was consequential not only for what was directly at stake, but for its effect on the Western position in other parts of the world."[16]

As long as the United States remains committed to the security of Israel and the West is dependent on Persian Gulf oil, the Middle East will remain important for American foreign policy. But, unlike the rivalry with the Soviet Union, it will not provide us with a rationale for a global foreign policy. As Mandelbaum says, "The Persian Gulf excepted, the United States is considerably less likely to dispatch forces abroad in the post–Cold War era. In this sense the gulf crisis belongs to the past, not the future, of American foreign policy."[17] There is, nonetheless, a need for a U.S. military force abroad, but it is in Europe and the Western Pacific, not in the Third World, where it will prove most useful. Above all, the mission of U.S. soldiers, sailors, and air force personnel in Europe and the Pacific will be radically different from what it has been in the past.

The new internationalism rests on the proposition that the age of superpowers is over. National power, after all, derives from a triad of military power, economic power and social cohesion.[18]

The Soviet Union only qualifies in the first instance, and the United States, while its economic power is still enormous, is constrained in its behavior by the limits placed on its ability to act by its dependence on foreign creditors. In moving from a creditor nation to the largest debtor nation in the world in less than a decade, the United States has shorn itself of superpower status. Superpowers are not dependent, as America was in the Persian Gulf, on other nations to pay for the wars that superpowers wage. As for social cohesion, the United States at the end of the twentieth century finds itself politically stalemated on the question of taxation for the public good. Its secondary education system is in shambles. It fails to provide necessary social services, such as universal healthcare, which European nations take for granted. It has failed to resolve the deep divisions among its people over racial issues. It lacks any solid consensus on a domestic program to restore the dwindling ability of the states and cities to come up with adequate public services.

Although Europe and Japan lack the military attributes that the U.S.S.R. and the United States have saddled themselves with, their economic prowess and technological sophistication would surely provide them with such a military capability, should they choose to obtain it. In this respect, the European Community (led by Germany) and Japan, along with the United States and the Soviet Union, are all great powers. In addition, China has the potential to become a fifth great power. Its technological advancement in the realm of nuclear weapons gives it a basis for further technological strides in the nonmilitary areas. The economic reforms of the 1980s were already propelling China into becoming a serious economic competitor in the global economy. If China can resolve the stalemate that has come about from an aging leadership's efforts to combine economic reform with political control, then that country will soon play a central role in the world community.

In this environment, in the absence of superpower status, what we see emerging at the end of the Cold War is a global balance of powers. Moreover, the United States, the Soviet Union, the European Community, Japan, and China are all unlikely to make war on one another. None of these states is an unsatisfied power, in the sense that it is bent on territorial expansion. We are seeing therefore a world likely to be free from the threat of great-power military conflict. The world has not yet realized Woodrow Wilson's vision of a universe of democratic nations, but, as Wilson's heir, Franklin D. Roosevelt, promised, it is a world that has been made relatively "safe for democracy." At this moment in history and probably for some time to come, the great powers share a common interest in preserving the peace, and to the extent that this characterizes the international system, limited collective security is possible, as we have seen in the Gulf war.

Collective security, in the strict sense of the term, means that an attack on one nation is an attack on all. This is not likely to be applicable in areas where the great powers are not active or where their interests are not directly affected. In much of Africa and in South America (though not in the Caribbean Basin or in Mexico), the great powers are unlikely to be involved. This does not imply indifference to conflict in these areas; the United States was active in resolving the Ethiopian civil war. It certainly implies a greater reliance on the mechanism of the United Nations to provide peacekeeping forces, to monitor free elections, and to supervise transition governments. But the exercise of U.N. functions will be possible only insofar as the great powers cooperate in this endeavor.

With the exception of Japan, the five power centers are all represented in the U.N. Security Council, and this will mean that the United Nations, as a collective security organization, will be limited to taking enforcement actions only against smaller powers; the threats to peace, moreover, are most likely to come from smaller powers. Should that situation change, it will mean that the great powers will once again seem to threaten one another, and the limited collective security that now appears possible will evaporate.

What we are talking about here is clearly not world government but a new experiment in international order, in which the three major economic powers—plus the Soviet Union, if it moves successfully toward a market economy and overcomes its drift toward disintegration, and China, once it begins to turn its economic potential into effective power—form a kind of central steering group, in which regional powers like India and Brazil will play comparable roles in their areas. But none of this can come into being without the leadership of the United States.[19]

As first among equals, the role of the United States is surely to sustain this global balance of powers. This will require an internationalist policy that will substantially reduce but not eliminate the U.S. military presence in Europe and Asia. In Europe, however, America's role will have changed. It will no longer be there to deter an expanding Soviet Union, but rather to reassure the Europeans that we are prepared to be a guarantor of the peace of Europe.[20] Germany, newly unified, will want to feel secure that its borders will not be threatened by renewed Soviet pressure. In their turn, the other European states, including the Soviet Union, will want to be reassured that Germany will not seek nuclear weapons.

Similarly, in East Asia and the Western Pacific, the United States has a role to fulfill by reassuring Japan that it will not be left alone to face pressures from China or the Soviet Union. Japan, like Germany, should not find itself compelled to acquire nuclear weapons for self-defense. Nor should it be encouraged to expand its non-nuclear military role, which would only alarm other Asian nations (not least China), which would assume that Japan was attempting to establish itself as a military as well as an economic power, a latter-day version of the East Asian Co-Prosperity Sphere that it espoused during World War II. The United States is essential to the regional balance of power that now exists in East Asia, but, as in Europe, it need not deploy the land, air, and naval forces it now does in the Philippines, South Korea, Japan, and the Western Pacific.

Historically it is virtually axiomatic that America's vital interests will always be located in North America and the Caribbean Basin. Canada and Mexico are among America's largest trading partners, and the policy of extending a free-trade zone to Mexico is a sensible recognition of this. But toward the region as a whole, especially Central America, the United States must be prepared to offer trade preferences and some measure of debt forgiveness. This, in turn, must be coupled with a policy of demilitarization of Central America. Like it or not, the region is considered a U.S. sphere of interest. Not only is the United States the natural market for the region, but it is directly affected by illegal immigration from the area and from political and economic strife that has led time and again to an exodus of political refugees. With the end of the Cold War, however, the United States can no longer use presumed Soviet expansion as a pretext for intervention. Yet paradoxically, in the post-Cold War environment, the United States will find its great-power role enhanced.

The central issue facing the United States, however, is to devise new economic institutions to replace or augment the existing international economic and financial system that was set up after World War II. Three great trading blocs with a tripartite monetary system threaten global economic stability: a European zone controlled by the D-mark; an East Asia dominated by the yen; and the Western Hemisphere under the hegemony of the dollar. The United States, which, with all its economic travails, still dominates the world economy and whose currency still serves as world money, must take the lead in devising new international economic and financial institutions before it is too late and the dollar loses its essential role in the global economy through sheer negligence.

To prevent the emergence of dangerously competing trading blocs that could easily lead the world into a Hobbesian struggle of beggar-thy-neighbor, the United States has the singular opportunity to devise a world central banking sytem and a new world currency to ensure monetary stability. A new international trade organization as a successor to the weakened General Agreement on Tariffs and Trade would be a natural complement to monetary reform and further help the now-emerging trading blocs to expand and adopt new rules on trade and investment.

But the new American internationalism will be stillborn if the United States does not accompany its concern with global security and international economic reform with a profound restructuring of the domestic economy. Unless the political stalemate is resolved in this country in a way that will allow Americans to enjoy the public goods that other advanced industrial nations take for granted, unless the twin trade and fiscal deficits are reduced or eliminated in order to provide a solvent basis for an effective foreign policy, unless political decisions are taken to turn Americans from dreams of unlimited expectations and endless consumption to investment and savings, then Americans will soon find that decisions will be made by others that will shape their lives and fortunes for good or ill.

America's place in the post–Cold War era may well seem more restrictive than the role the United States played as a superpower challenging the Soviet Union for global supremacy, its national mission for over four decades. But the great lesson of how the Cold War ended, as columnist William Pfaff reminds us, may be expressed in these words: "The Soviet system collapsed because of what it was, or more exactly, because of what it was not. The West 'won' because of what the democracies were—because they were free, prosperous and successful, because they did justice, or convincingly tried to do so."[21]

Ultimately, the success of the United States will lie in the quality of its society. Under these circumstances, as I hope to demonstrate in the pages that follow, an American commitment to working out new principles for limited collective security by acting to sustain a global balance of powers, and by revising outdated economic, political, and military arrangements, is hardly a retreat from power. It is rather a recognition that the end of the Cold War requires a new internationalism to take account of our interests in the years to come. As Walter Lippmann wrote in the midst of World War II, "when we know what we ourselves need and how we must achieve it, we shall be not only a great power. We shall become at last a mature power. We shall know our interests and what they require of us. We shall know our limitations and our place in the scheme of things."[22]

PART I

Questions
of Solvency

1

Insolvent America

W<small>HEN</small> the American security system—stretching from the Elbe River between the two Germanies to the Yellow Sea—was set up in the early 1950s, the United States was the most powerful nation in the world. It accounted for over 40 percent of the gross world product. It produced 30 percent of world manufacturing exports. In the 1990s America produces about 20 percent of the gross world product. America's rate of productivity, once the greatest in the world, fell behind that of Japan and West Germany and some of the newly industrializing countries on the Pacific rim of Asia.

Throughout the postwar era—and especially up to the mid-1970s—the dollar was the keystone of the American economic position and the central mechanism for world monetary stability. In the words of Princeton political economist Robert Gilpin, it "cemented the American system of global alliances and has been the foundation of American hegemony."[1] With the dollar as the basis of the world monetary system, we have been able to fight foreign wars, station troops in Europe, Asia, and Latin America, and build a navy that could patrol the world's oceans. And all this has been done without putting a heavy burden on the American taxpayer. During the Vietnam War we printed money in order to fight it without paying new taxes; during the Reagan and Bush administrations we borrowed from foreigners so heavily to finance federal expenditures for defense, education, and other programs that our foreign debt came to total about half a trillion dollars by 1991. (A decade earlier, we were a net creditor with an official position of positive $141 billion.) Our net debt may well reach $1 trillion by the mid-1990s.[2] Dividends and interest on this foreign debt cost American taxpayers between $40 and $50 billion a year.

To pay off this debt, the United States has to create productive assets, as it did prior to World War I. The problem is that during the Reagan years capital from abroad largely financed private and public consumption, including the military buildup, rather than domestic investment. While military spending can provide a positive stimulus to the economy through spin-offs, such as computers, radar, and jet aircraft, these technological advances could have been achieved more efficiently through direct research into these areas. As a consequence of Ronald Reagan's and George Bush's policies, an indebted citizenry will have to undergo a substantial lowering of its living standard. As it is, the average annual growth in per capita gross national product (GNP) has fallen to –0.4 percent during the first three years of the Bush administration, the worst record of any president in the post-World War II era. Both future consumption and future investment are likely to suffer.[3]

America's profligate ways grew at an alarming rate during the two terms of Reagan's presidency. The federal budget deficit rose from $79 billion in FY 1981 to a projected $282.3 billion ten years later.[4] Yet the Reagan administration, as did the Bush administration, came to office promising to balance the budget and so set the United States on the road to solvency. In order to maintain our commitments abroad, we were supposed to earn enough to sustain them. Under the "Reagan Revolution," taxes were to fall and productivity was supposed to rise; and Americans were to put their new savings into the domestic economy.

But it did not work out that way, and the American consumer went on a buying spree. At the same time both the Reagan and Bush administrations embarked on a policy of increasing the defense budget to record levels not seen in peacetime without ever asking the citizens of the country to pay for them.[5] While the administration and the Congress were finally forced to show some movement in reducing the federal deficit, the legacy of past profligacy ran high and no great sacrifices were demanded. The foremost motive of those involved in trying to fashion a deal over the budget in the summer and fall of 1990 was to avoid a prolonged public debate and political conflict over the basic issues that underlay the crisis.

The other tower of debt—the trade deficit—reached $152.1 billion in 1987. By driving down the value of the dollar, the administration managed to reduce the deficit to about $118 billion by the end of Reagan's term in office.[6] This meant that the United States would still have to achieve an improvement of about $200 billion in its trade position, because the 1988 deficit in current account (which represents trade in goods, services, and investment) was running to about $150 billion (compared to a surplus of $6.9 billion in 1986), and debt-service costs were expected to rise by about $50 billion by the early 1990s. Although the trade deficit dropped to $101.1 billion in 1990, the improvement came almost entirely in imports, reflecting a weak economy.[7]

How then have we financed our deficits? With relatively low inflation rates, we have not yet started printing money to pay for them. Instead, we have used our relatively high interest rates to attract foreign capital. In a word, those obliging foreigners allow the American consumer to buy his or her foreign-made television sets and VCRs, and with the dollars that flow abroad, foreigners not only buy up U.S. Treasury offerings that underwrite the deficit but also acquire office buildings and real estate. Americans in their passion for consumption are selling off their patrimony.

In essence, here is the causal relationship between the federal budget deficit and the trade deficit: to finance the budget deficit, you need to raise interest rates. This results in a capital inflow and an overvalued dollar. The high dollar, in turn, means a decrease in U.S. exports and a corresponding increase in imports. Hence the trade deficit.[8]

The need for a public reckoning seems very much in order, but there is no way this can be done unless the American people reexamine their defense commitments in the light of their financial obligations. The United States, after all, maintains a military establishment that will consume a projected $291 billion in FY 1992 despite the estimated budget deficit in 1991 of $318 billion.[9] After adjusting for inflation, the military budget initially requested by the Bush administration for 1991 came to $306 billion—30 percent higher (adjusted for inflation) than the 1980 defense budget before Reagan began his massive military outputs.

But the Soviet Union is no longer a global threat that must be countered with a policy of global containment. The most serious threats to American security are the dangers at home, where a deteriorating infrastructure, a desperate underclass, a degenerating environment, an inadequate educational system, and an indebted citizenry render the body politic singularly vulnerable to domestic upheaval as America seeks to play a constructive role in the coming century. Under these circumstances, we should establish priorities at home of the same magnitude as the statesmen did in the immediate postwar era to cope with the exigencies of the Cold War.[10]

The costs of excessive military power to the domestic health of the nation are severe. Although the defense budget takes up a little over a quarter of the federal budget, a large part of that budget consists of self-financing programs, such as social security; defense, therefore, takes up more than half of each income tax dollar. As for research and development, the military share of federal R&D is between 65 and 73 percent, and the military share of all American R&D is between 30 and 43 percent; yet defense R&D no longer drives commercial innovation. There is, in fact, growing evidence that it retards commmercial innovation by competing for scarce, technologically trained manpower. No comparable diversion of R&D from the civilian to the military sector can be found in Japan and Western Germany, our main economic competitors, who spend respectively only 4.5 percent and 12.5 percent of their R&D on military matters.[11]

More worrisome still is the deterioration of the infrastructure of the nation. It will probably take $51.4 billion just to replace or rehabilitate all the deficient bridges in the country. It will cost more than $20 billion to repair deteriorating public housing. Eliminating hazardous waste on military bases may cost $14 billion. Modernization of the increasingly hazardous air traffic system could run $25 billion. Yet total public spending on infrastructure dropped from 3.6 percent of GNP in 1960 to 2.6 percent in 1985.[12]

As for the environment, scientists now warn of a so-called greenhouse effect caused by pollutants in the atmosphere. A gradual warming trend in the temperate zones could produce severe drought conditions. Because the greenhouse effect results from a buildup of carbon dioxide from the burning of fossil fuels such as coal and oil and other gases, substitute fuels must be found to slow or halt the pace of global warming.[13]

To cope with these problems alone will surely require Washington to reexamine not only its levels of taxation and the need to encourage investment at home but also the extent and costs—both tangible and intangible—of its security. In this respect, a severe cut in the defense budget is the logical conclusion America should reach as it seeks to provide for the well-being of its citizens. If security is no longer to be found mostly in guns, ships, and aircraft, then there is no reason to spend half the nation's budget on defense. Such spending could drop by some $150 billion by the end of the decade, and this would still allow the United States to play a global role in international affairs.

To achieve this level of defense expenditures would require a two-pronged attack: a dramatic reduction of conventional forces and a radical slowdown of the modernization of nuclear weapons. Both our defense buildup during the Reagan years and the debates over the size of the budget in the Bush administration bear little relevance to any overall strategy. The Reagan buildup resembled at times a Chinese menu of men and munitions: faced with column A and column B, secretary of defense Caspar Weinberger seemed unable to make choices based on need, and thus chose everything. After enormous reluctance, Bush's Secretary of Defense, Dick Cheney, finally had to recognize the need to cut back on defense spending when the Soviet Union was itself cutting back on all fronts. But even with the projected cuts in U.S. forces serving in Europe by 50 percent by the end of the decade, there will be little savings in the defense budget unless the troops are not only brought home but also demobilized. Once again, little thought seems to have been given to the actual defense needs of the United States in a world turned upside down. Even when Bush announced in September 1991 that the United States would unilaterally eliminate all tactical nuclear weapons on land and at sea, and halt further development of some long-range nuclear weapons like the 10-warhead MX missile, the president warned that this would not result in any cuts in the defense budget. By urging continued funding of experimental weapons systems like the B-2 Stealth bomber and the Strategic Defense Initiative ("Star Wars"), Bush's defense budgets might actually rise.[14]

Planners outside the Pentagon, however, have begun to think about the demands of the post-Cold War world. The Brookings Institution produced a report in the fall of 1991 concluding that America's defense spending can be reduced by more than a third over the next 10 years, and still America would maintain the world's preeminent military establishment. The report estimated that cumulative savings could be realized of some $316 billion, and up to $619 billion if production costs of a new generation of high-technology weapons expand (as predicted by the Congressional Budget Office). The authors of the Brookings study point out that "the view from the Pentagon is of a scaled-down version of the Cold War."[15]

The conflict in the Persian Gulf over Iraq's seizure of Kuwait, which resulted in the most serious buildup of U.S. forces since the Vietnam War, only reinforces the argument for a new strategy to deal with a world shorn of superpower rivalry. In confronting Saddam Hussein, the American forces did indeed find themselves short of equipment—not glamorous weapons like the B-2 Stealth bomber, but more mundane weapons the Pentagon chose not to buy during the halcyon Reagan years. The Navy, for example, found that it had so few fast transports that it could move only one division at a time to the Middle East. Minesweepers, another humble craft, have never been high on the Navy's most-wanted list, yet mines are especially effective in narrow waterways. When an Iraqi tanker was suspected of laying mines in the Persian Gulf during the Iraqi invasion of Kuwait, the Navy had three aircraft carriers in the region but no minesweepers. Repelling attacks on U.S. forces in the Middle East, for example, will require not the use of the B-2 bomber, but the A-10, a tank-busting plane the Air Force considered so slow, cheap, and unexciting that it closed down its production line.[16]

The collapse of the Soviet empire in Eastern Europe is surely convincing evidence that we will not see a large-scale conventional war in Europe in the years ahead. Nor will we see the "limited nuclear war" that nuclear theorists have long been predicting. What we are likely to witness is greater bloodshed at home by the importation and use of narcotics, more lives lost as international terrorists become better armed and more desperate, and the disruption and destruction of the world's vital natural resources by criminally irresponsible governments. The Iraq–Iran war, the subsequent Iraqi invasion of Kuwait, and the turmoil in Central America and the Philippines indicate the likelihood that U.S. armed forces will be used—if at all—in regional conflicts, not in a great land war in Europe or the Far East. How and whether we restructure our armed forces and rethink our strategy to meet these contingencies will be the central military question of the next decade.[17]

Beyond cutting the defense budget, there are any number of ways that the United States could restore its economic and financial well-being, even though we can never recover the all-powerful role we enjoyed in the immediate postwar era. We could tax gasoline at the pumps, tilt tax policy to favor investment and savings, tax social security at a higher rate for couples who are well off, and cut cost-of-living adjustments in half until federal retirees reach their sixty-second birthday; today many such retirees now receive full pensions as soon as they leave government service—for many, in their early forties. We could close loopholes in the current law that allows as much as three-quarters of capital gains to go untaxed. We could tax the affluent at roughly the same rate as the less-well-off, and close loopholes in the corporate income tax laws.[18] The United States, compared with the affluent nations of Western Europe, is undertaxed.

But even if Congress and the administration showed the political will to make these adjustments, their aim should be not only to narrow the budget deficit but also to invest in America. It is one thing to borrow money for new weapons systems and for our inadequate social services, and another to rebuild cities and improve education and transportation—in short, to use deficit spending to promote economic growth in the very sinews of the American economy rather than to finance public consumption under the illusion that we live in a world of limitless expectations.

There is a hard reckoning for an insolvent foreign policy. In the absence of the Soviet threat, the United States will no longer retain its traditional leverage over the Europeans and the Japanese as their protector against a Russian attack.[19] Americans will simply have to learn to live with the political impact of the power of Japan and a Europe of 700 million people in a world where the primary threats to instability will be economic. Even military threats to Western security are most likely to arise when access to natural resources is affected, and hence the stability of the global economy is endangered. In these instances, the United States should be able to take a leading position without impoverishing itself.

The role America can play in reassuring the Europeans that the post–Cold War peace will endure, America's place in East Asia as the holder of the balance of power, its efforts to construct a free-trade zone in North America, its ability to respond to troubles that affect its vital interests elsewhere, its ability to take a leading role in devising new international financial and economic institutions to promote global prosperity—all these roles and responsibilties can be assumed only by a solvent America that brings into balance its commitments and its capabilities.

2

The Second Russian Revolution

THE former Soviet Union was hardly a mirror image of the United States. In 1989, the U.S. gross national product totaled about $5.2 trillion; the Soviet Union's officially came to only $2.5 trillion, less than half of America's—and it was doubtless even less. With an average growth rate of less than 2 percent in the 1980s, the Soviet Union is falling behind Japan as well as the United States, and at this rate it may well fall behind China in twenty years.[1] With the challenge of developing new technologies, the Soviet Union could no longer afford simultaneous increases in consumption, civilian investment, and military production. And so, faced with the desperate need for relief from military and technological competition with the West, Mikhail Gorbachev rose out of the Soviet bureaucracy in 1985 with a program of *perestroika*—or restructuring—of the Soviet economy. Eager to save socialism by guided, if radical, reforms, Gorbachev found that he had unleashed a revolution. Five years later, not only was *perestroika* threatened, but the very existence of a centralized Soviet state.

In August 1991, the unintended consequences of Gorbachev's reforms came to a dramatic head. A coup to replace Gorbachev with a more cautious leadership was

29

mounted by conservative elements of the military, the KGB secret police, and the Communist party bureaucracy. Its failure was a world-historical event of momentous proportions.[2] In its wake, the Communist party was dismantled, and an economic directorate was put together to speed the Soviet Union on the path to a nonsocialist economy. At the same time, the centrifugal forces that had been pulling at the Soviet Union grew far stronger. The Baltic States—Lativia, Lithuania, and Estonia— were granted independence by Moscow. Nine other republics also declared or pushed for independence. The likelihood was a dissolution of the Soviet empire—at least as the world had known it since the end of World War II.

Moreover, the legitimacy of the Second Russian Revolution derived from the people who supported the president of the Russian Republic, Boris Yeltsin, in his insistence that the legally constituted president of the Soviet Union, Mikhail Gorbachev, be returned to power. George F. Kennan, the leading American expert on the Soviet Union, asserted that the events surrounding the failed coup actually eclipsed the 1917 Russian Revolution in importance. "For the first time in this history," he said, "they [Russian people] turned their back on the manner in which they've been ruled—not just in the Soviet period but in the centuries before. They have demanded a voice in the designing of their own society."[3]

The need for the Soviet Union to prepare to enter the twenty-first century not simply as a superpower in decline but at the very least holding its own among the other great economic powers in both East and West has been the central aim of the current generation in power. Only an economy, in Gorbachev's words, "developing on the basis of a state-of-the-art scientific-technical base . . . will deservedly enter the new century as a great and prospering power."[4] Thus, the leadership was humiliated by the breakdown of the Soviet economy. Today, the Soviet Union remains a military superpower; otherwise, its cities resemble those of a middle-rank developing country, and its countryside is mired in backwardness and shocking poverty.

To ask whether Gorbachev and Yeltsin or their immediate successors will succeed or not in their call for economic reform is to pose the wrong question. The process can hardly be reversed, but what Gorbachev intended when he took power, the "liberalization" of the Soviet economy, has now given way to a far more radical approach toward achieving a market economy. To get there, however, may be beyond the reach of the Soviets in this generation. The Russians, after all, have almost no notion of what a market economy is. As Anatoli Adamislim, a Soviet deputy foreign minister, quipped: "To alter the world is difficult; to alter human beings is almost impossible."[5] The Russians think of the vital middlemen as mere speculators. Yet no one is likely to be able to offer clear alternatives.

In the first five years after Gorbachev took power, his economic policies made everyday conditions worse. Although the men in the Kremlin recognized the crisis, they themselves were astounded at its depth. By 1989, the Soviet economy was showing no industrial growth. A year later, industrial production was actually falling; for 1992, the Soviet GNP was projected to decline by 13 percent.[6] International trade figures showed a $5.4 billion deficit for 1989—the first unfavorable balance in fourteen years. By 1991, current Soviet foreign debt totaled $70 billion. The only thing in plentiful supply was cash, but there was little that was worth buying, certainly not the Soviet television sets that tended to catch on fire or the shoes that had the durability of cardboard. The Soviet Union's citizens are about as affluent as Portugal's, but unlike the Portuguese they can find nothing worth buying.[7] Barnstorming in the Urals to gain support for radical changes in the economy, Gorbachev pointed out: "If we do not get out of the system we're in—excuse my rough talk—then everything living in our society will die."[8]

In essence, Gorbachev tried throughout the late 1980s to set in place piecemeal reforms, but without any precise idea of what he wanted to achieve. Although the Soviet leader was probably fearful that the Soviet Union could not withstand the shock therapy of plunging headlong into a free-market economy, he offered no real alternative. Like a man lost in the dark and feeling his way to safety, he only slowly discovered what his chief economic advisers were urging—to move toward various free-market institutions, including variable prices, a convertible ruble, an independent commercial banking system, and a stock market in which enterprises would finally be free of the central dictates of the state planning ministries and put into the hands of shareholders.[9] Yet until the aftermath of the August 1991 coup, he hesitated to embrace a full transition to a nonsocialist economy.

Though there are deep questions as to how much autonomy will be wrested by the republics from the union government, economic planners in both spheres agree on the need to commence a massive program of privatization, which means turning over the assets from the state to private hands. They envisage price liberalization, monetary and financial reform, and cuts in government spending. After the coup, the leaders of eight republics agreed on the need for an economic community; however, unless such a grouping were to have a unified monetary policy, a common currency, and a role for the center in tax policies, the Soviet economy is likely to run out of control on a wave of inflationary spending.

On the other hand, if successful, converting the Soviet Union to a market economy might make it possible for that country to enter the twenty-first century, in Gorbachev's words, as "a profoundly democratic state, its economy an important and integral part of the new global economy."[10] But what Gorbachev did by opening up Soviet society and attempting to restructure it was to risk the destruction of the Soviet Union as such, because *glasnost* and *perestroika* did not so much save society as challenge it. Gorbachev had, in a sense, radicalized it in a democratic direction and shifted the balance from a moribund ideology toward a new creed of citizenship, what he himself once called "a democratic country based on the rule of law, a real civil society."[11]

In fact, there was always more of a civil society under the rule of law in the late tsarist period than was generally evident to outsiders; yet while a civil society can lead to real democracy, these aspirations are likely to be fulfilled only if the economic situation within the Soviet Union reveals a change for the better. The great danger is that as the Soviet leadership tries to cope with a disintegrating internal empire, it will be supplanted by a reactionary nationalist government supported by a disaffected military.

It was precisely to forestall the transition to a civil society and to insist on the territorial integrity of the Union that the plotters planned their putsch. Gorbachev had already rejected the so-called 500-Day Plan in the fall of 1990 when he came to realize that by fully introducing a market system and privatizing state enterprises he would inevitably destroy the Communist party's monopoly on power. This was something that Gorbachev could not bring himself to do before the August coup.[12]

In addition, Gorbachev understood that the plan's recommendations to devolve economic power to the republics would result in the dismemberment of the Union. His fears were echoed by the Soviet apparat in the party, the army, the KGB, and the military-industrial complex (responsible for half the country's economy and probably employing one-third of its work force). Gorbachev's original strategy of reform communism, however, was failing, and yet the Soviet president seemed to have no better solution than tactical maneuverings between the forces of the right and the left, hoping that an infusion of foreign aid would somehow allow him—and the Union—to muddle through.

The plotters, for all their inept planning, were not a gang of right-wing reactionaries. On the contrary, the leaders came from Gorbachev's own cabinet, and they were all members of the central institutions of the Soviet system. Their decision to try to overthrow Gorbachev was probably made sometime after April 23, when Gorbachev capitulated to the demands of Boris Yeltsin to accept the idea of genuine autonomy for the republics. The new Union treaty was to be signed on August 20; two days before that date, Gorbachev was put under house arrest while vacationing in the Crimea.

Yeltsin's success in turning back the coup, however, was not due only to the staggering incompetence of the plotters; he achieved his aims because he was able to rally soldiers, workers, and even members of the KGB to his calls for legitimacy. In the words of the American historian Martin Malia, "what began as a coup within the Party was transformed under the new democratic conditions created by a revived Russian civil society into a genuine and world-historical revolution."[13]

A communist regime that, as Czechoslovakia's president Vaclav Havel once said, "justifiably gave the world nightmares," came to an end.[14] The future of what remains of the Soviet Union now lies in the hands of the people, who will demand universal and direct suffrage, the only basis in the modern world for democratic legitimacy.

For Americans, the central issue is the security of the West. Should Gorbachev and Yeltsin disappear from the scene, would this be threatened? The Soviet Union has already effectively withdrawn from Central Europe and the Balkans. When Gorbachev said in October 1989 that the U.S.S.R. had "no right, moral or political," to interfere in events happening in Eastern Europe, he reversed the course of empire that had informed Soviet policy since the end of World War II. Arms control negotiations between Washington and Moscow that had been stalled for years were lent new energy by the Kremlin's unilateral announcements of dramatic drawdowns of Soviet troop strength and weaponry in Europe. The revolution started by Gorbachev affected every dimension of Soviet foreign policy. Not only are the Soviets withdrawing from the center of Europe, but they have also negotiated with the Chinese for troop reductions in East Asia, and have made a series of proposals to lower tensions in the Western Pacific by suggesting lower levels of naval deployments.

Toward the Third World, where the Cold War has been fought out since the 1960s, Moscow also pulled back its horns: it helped negotiate a settlement in Namibia, put its Middle Eastern friends on notice that they could no longer expect automatic support from the Soviet Union in any confrontation with Israel, lined up with the West in condemning its erstwhile ally, Iraq's Saddam Hussein, for his invasion of Kuwait; worked to persuade its Vietnamese allies to withdraw from Cambodia in order to further chances for peace; and cut its aid to the Marxist forces in Central America and Cuba. More important than all these initiatives was the decision by the Soviet leadership to withdraw Soviet forces from the debilitating and humiliating war in Afghanistan—what Gorbachev called the "bleeding wound" in the Soviet body politic. After 1986, as the American Sovietologist Robert Legvold noted, the Soviet leadership appeared to view the Third World as a "tragic arena, not a region of hope and promise."[15]

The once-hallowed precept that the struggle between capitalism and socialism formed the central dynamic of international politics no longer holds sway in the U.S.S.R. Instead, the notion of entanglement and cooperation rather than confrontation and conflict is the likely axiom of the post–Cold War world. One of Gorbachev's early advisers, Aleksandr Yakovlev, has spoken of a "planet compressed to an unprecedentedly small size"—a world needing not the primacy of "individual countries or classes, peoples, or social groups," but ways of countering "the forces of separation, of opposition, of confrontation, and of war, which have already delayed the development of civilization by whole centuries."[16]

In short, almost no one in either the United States or the Soviet Union believes that turmoil in Asia, Africa, the Middle East, or Central and South America is part of some grand liberation struggle by the forces of socialism against the imperialistic designs of the capitalist West, an excuse for Soviet commitments and intervention. Instead, it is seen as a drain on Soviet resources as well as on the resources of the developing countries themselves. As Gorbachev declared in his speech to the U.N. General Assembly in 1988, echoing John Donne: "The bell of every regional conflict tolls for all of us."[17]

The revolution in Soviet foreign policy that has led the Soviet Union to reassess its commitments and its capabilities is in part a function of its domestic problems: how can the U.S.S.R. enter the twentieth century as a superpopwer if it does not modernize its economy? Just as Charles de Gaulle insisted that the French army withdraw from its colonial entanglements to become a modern army able to compete with, if not surpass, the two superpowers, so, too, Gorbachev and his military advisers understood that the strongest argument for radical economic reform was that technological backwardness would be dangerous for the security of the Soviet Union. But domestic travail alone was not decisive in turning Soviet foreign policy to a new course. The Soviet leadership also recognized that the Soviet Union cannot isolate itself from the global economic, social, and environmental forces that affect all nations, and that inevitably forces a new definition of national security.

Despite Gorbachev's recognition that the U.S.S.R. was overextended, he never intended to abandon Eastern Europe in such a hasty and unseemly fashion. He certainly never expected to witness the disintegration of the Soviet alliance system and the possible loss of valuable buffer states. Yet these were the unintended consequences of a policy that came to favor West European integration, encouraged reformism in Eastern Europe, and repudiated the Brezhnev Doctrine, which had demanded ideological purity in the Soviet bloc at the risk of direct Soviet military intervention. The last thing Gorbachev anticipated was German unification.

History, however, does not go in reverse, even if there are those like King Canute who would like to turn the tide backward. At best, reactionaries can build dikes against a new tidal wave; but even these dikes last only a short time before they too are overwhelmed by the forces that have already been released. Gorbachev or Yeltsin—or any other Soviet leader—will have to adjust to the altered world we now live in. Invoking a foreign threat no longer furthers the domestic goals of the Soviet Union. "The main threats to the security of our society now emanate not from outside, but from within," wrote Andrei Kortunov and Alexei Izumov, foreign policy analysts from Moscow's USA and Canada Institute, "and are connected (in diminishing order of importance) with economic collapse, the nationality crisis, the deterioration in the environment, and the growth of crime."[18]

It was the Gorbachev generation, not just the man, that came to power in the mid-1980s, and while there are often sharp differences in the tactics that might be pursued to make the U.S.S.R. technologically competitive, there is a consensus among the ruling Soviet elites that this cannot be done without shedding much of the baggage of Soviet foreign policy that proved so financially burdensome and politically unrewarding in the past.[19]

Yet the Soviet Union is still a great power. Whether it can remain so is another matter. In the 1990s, the Soviet Union finds itself without significant allies and with uncertain frontiers, an economy in crisis, many of its peoples alienated, and facing the wholesale disintegration of its federation. This is not a recipe for greatness. Looking outward, the Soviet leaders are being forced to rethink some of the basic premises of their foreign policy. Should it seek a formal military alliance with the West? Should it withdraw into sullen isolation?

No matter who succeeds Gorbachev, the Soviet Union would find it impossible to reoccupy Eastern Europe without risking a global conflict. Until the economy shows real signs of providing the citizens of the Soviet Union with a decent measure of well-being and until some minimum degree of national unity is achieved among the non-Russian peoples, the great preoccupation of the Soviet leadership will necessarily be with domestic affairs. Even if a nationalist coalition, backed by disaffected elements of the military, were to come to power, this kind of leadership would provoke such a hostile reaction by the non-Russian parts of the Soviet Union that the Soviet leadership would be faced with civil strife that would virtually paralyze the Soviet state; it would certainly absorb any remaining energies that the Soviet leaders might like to devote to international affairs. It would virtually preclude military intervention abroad.[20]

A more promising future would see a Soviet leadership that coped with the problem of retrenchment by strengthening international institutions. In this respect, for the first time in a generation the United Nations may begin to fulfill the role it was designed for. The founders of the United Nations knew the organization could not work effectively without the cooperation of the five permanent members of the Security Council. Yet it was not until the late 1980s that the five permanent members began to work together to seek an end to conflict—for example, the Iran-Iraq war and in southern Africa—and to discuss the dangers in the Persian Gulf, not only in official meetings but, as I was told by a member of this group, "in their own living rooms." None of this could have taken place without the acquiescence of the Soviet Union.

The first significant indication of the Soviet Union's turning away from its traditional ideological struggle in the Third World and its new emphasis on resolving regional conflicts through the United Nations was a 1987 article in *Pravda* and *Izvestia* by Mikhail Gorbachev calling for a "comprehensive system of international security."[21] At that time he even went so far as to suggest that the Security Council's permanent members "become guarantors of regional security." The old Soviet approach to global issues had been narrow and legalistic: everything not specifically in the United Nations charter was to be brought before the U.N. only at the sufferance of the Soviet Union and the other permanent members of the Security Council.[22] The new approach was broad and inclusive.

Now Moscow accords the United Nations the right to deal with questions ranging from the peaceful resolution of disputes to the carrying out of human rights agreements, from the transfer of conventional arms to space exploration and the poisoning of the earth's oceans. Above all, "comprehensive security" assumes an interdependence among nations, an idea that Moscow once saw as a way the West kept down the poorer countries of the Third World; now, the fact of interdependence is viewed as the most logical approach to the nuclear and ecological threats to the planet."[23]

Gorbachev's emphasis in his 1988 speech at the United Nations was to de-ideologize foreign policy. It was in this light that he called for the U.N. secretary general to help untie, as he put it, "the knots of regional problems." Along these lines Gorbachev suggested that countries routinely assign some of their armed forces for U.N. peacekeeping duties, which would mean giving the United Nations a kind of army of its own. (He also recommended this for the Gulf war.) Specifically, Moscow urged the United Nations to expand its peacekeeping operation in order to enforce decisions of the Security Council and "for the prevention of emerging armed conflicts." By being willing to provide Soviet troops for such duties, Gorbachev offered to change the informal rules that preclude the great powers from taking part in peacekeeping operations."[24] After Iraq's invasion of Kuwait, Gorbachev repeatedly called on the United Nations to sanction any military intervention against Saddam Hussein.

Another startling Soviet proposal involved the International Court of Justice (the so-called World Court), which Gorbachev suggested could be used by the Security Council and the General Assembly for "consultative conclusions on international disputes," with its "mandatory jurisdiction recognized by all."[25] Such a system implies at the very least a shift away from a cynical use of the General Assembly, which ordinarily reflects Third World interests. More important, for the first time the Soviet Union was not standing steadfast for all the privileges of a sovereign nation but rather seemed willing to engage in a process involving laws and international institutions that would restrict its freedom of action—a stunning concept for a communist state.

Despite all this emphasis on internationalism and interdependence, the vision that Gorbachev evoked still held the great powers as the managers of an inherently anarchic world. His departure from Marxist doctrine therefore proceeded in two directions. His internationalism was made up of global issues—

such as environmentalism, the exploration of space, and the seabeds—that affect everybody. To deal with these problems, however, he returned to the system of great powers reminiscent of traditional Russian statecraft as practiced by Peter the Great.

In the near term, then, we are still bound by considerations of the balance of power. This is especially relevant to international order after the Cold War, for it is the global balance of great powers that now makes it possible to take collective action to achieve settlements in those parts of the world in which the great powers are involved. And in no region has the balance of power changed more decisively than on the European continent after the precipitate. withdrawal of the Soviet Union from Eastern Europe and the subsequent reunification of Germany.

3

The Collapse of the Old Order

In reflecting on the momentous events that changed the face of Europe in 1989, what is surely the most remarkable is the rapidity of the revolution that swept westward from Poland. While the revolution could not have succeeded without the Soviet Union's willingness to withhold force, it is also striking that these regimes were so fragile. Nationalism in Eastern Europe, however, was far from fragile, and the governing parties had little if any authenticity. What shreds of legitimacy they might have possessed had been stripped way by their economic performance. But perhaps most telling was the sense of injustice felt by the workers to whom so much had been promised. In Czechoslovakia and in Hungary, the revolution may have been ignited by the intellectuals, but it succeeded because of the workers.

Not only did the Russians accept the collapse of empire, they unwittingly contributed to its collapse. Gorbachev's espousal of reform communism—his *perestroika*—validated the economic reality the East Europeans were confronted with: that their standard of living was falling so fast and so far that soon they would be considered little better than Third World nations.

In addition, the consequences of *glasnost* and *perestroika* led to a loosening up of the Soviet Union's imperial ties to Eastern Europe. Ending the Cold War was simply not possible without a basic reordering of the Soviet relationship to Eastern Europe.

For Gorbachev, his belief that the threat from the United States was greatly exaggerated inevitably led to his conclusion that the military and economic commitments needed to maintain Soviet power in the face of a largely imagined threat was devastating to his country's economic development and, in the long run, untenable. Confrontation and *perestroika* simply could not be combined.[1]

By offering to make significant cuts in troop strength in Eastern Europe, by signaling an end to the Brezhnev Doctrine that justified the right of the U.S.S.R. to intervene in the internal affairs of Soviet-bloc nations "when socialism in that country and the socialist community as a whole is threatened," Gorbachev signaled not only his willingness to tolerate a broad definition of what constitutes socialism, but also his understanding of what the United States required if the Cold War were truly to be put to rest. The Cold War might be said to have begun over Soviet unwillingness to hold free elections in Poland, which presaged the later division of Europe into two blocs; it could not therefore be ended without allowing the subject populations freely to have their say without the threat of Soviet intervention. Once this happened, even conservative skeptics in Washington had to accept the reality of the peace.

At the end of 1989, when presidents Bush and Gorbachev met at Malta, the very island where Churchill and Roosevelt held a rendezvous before their meeting with Stalin at Yalta, it was in a very real sense a meeting to ratify the ending of the Cold War. Bush stated forthrightly that "we stand at the threshold of a brand-new era of U.S.-Soviet relations." And Gorbachev echoed this sentiment: "The world leaves one epoch of the Cold War, and enters another epoch."[2]

There were, of course, a number of political events in one socialist country that affected the politics of another. And the speed of communications, in particular the access to television, kept the fuse of revolution burning. Yet it seems fair to say that events in Poland ignited the emancipation of Eastern Europe. Even though Solidarity had been suppressed in 1981, the movement was not crushed. The wave of strikes that lasted through the summer of 1988, in response to the economic devastation the communists had brought about, led to legalization of the trade union movement and finally to national elections in 1989, in which Solidarity would participate. When these elections were finally held in April 1989, the communists suffered a crushing defeat—in an election that had been skewed in the communists' favor by an election law that had been agreed to by Solidarity. Soon after, a Solidarity-led government took over.

All this transpired, let us not forget, without even the threat of Soviet intervention, something that would have been unthinkable eight years earlier. Then, in the fall of 1989, Hungary opened its border to East Germans seeking to cross into Austria en route to West Germany. The East German regime fell soon after, the Berlin Wall came down, and then came Czechoslovakia's "velvet revolution," and then came the Bulgarians' turn, and then, in a torrent of blood, the Rumanian coup.

For Gorbachev, in particular, the admitted failure of socialism in Eastern Europe was a severe blow for which he had not been prepared. When he first took over as party chief in 1985, Gorbachev apparently believed that prospects for economic improvement in Eastern Europe were far more promising than those in the Soviet Union. Except in Poland, the regimes seemed highly stable. At a meeting with the economic secretaries of the East European Communist parties in November 1984, he told them that he knew they were deeply troubled over their economic difficulties. "Some of you think you've found a life preserver in the 'market,'" he said. "But you should think first of all not about the life preserver but about the ship, and the ship is socialism."[5] Reformed socialism, combined with greater Soviet attention to East European sensibilities, would surely prevent any upheavals in the Soviet bloc. Yet in early 1988, there was a crisis of authority in the Hungarian party and Janos Kadar, the inventor of "goulash communism" and ruler of Hungary since the 1956 Revolt, was forced to resign. Not long afterward, a second series of strikes led to a legalized Solidarity party and the preparation for free elections. With these two developments, the signals were being picked up by the other East European parties that the Soviet Union was ready to significantly broaden its definition of reform communism.

Some of the Communist parties still thought they had found a formula for survival in a liberalized communism—but it was evident in Hungary and Poland that this brand of communism simply did not work. It did not provide a significantly better economic environment, and it was no longer seen as a practical alternative. While Gorbachev himself still hoped for guided reform within the context of liberalized communism, he was growing increasingly impatient with the stubborn hardliners in Czechoslovakia and East Germany, and at the same time was prepared to let them go their own way. The East European elites had always desired more independence from Moscow, and now they were getting it. But paradoxically, while they gained more freedom to do what they chose, they were losing the guarantee that Moscow would keep them in power; until Gorbachev, the Russians would sometimes fire the party chieftains but they would always preserve the party power structure. The Soviets, at this time, refused to bail out the East European communist regimes, and this, too, led to unintended consequences for the U.S.S.R.

By tolerating East European diversity, Gorbachev lost significant support among conservatives, and he stimulated the restive nationalities within the Soviet Union to demand their own brand of freedom. From a strategic viewpoint, the Soviet Union lost the buffer zone that had long served to protect it from invasion from the West. Moreover, this led to the unification of Germany. Although Gorbachev insisted that the U.S.S.R.'s external security could best be found within a global system of mutual security, what the Soviet "right" perceived was that Russia had lost its most tangible territorial gain in this century—the glacis of Eastern Europe. On the "left," however, the Russian intelligentsia felt vindicated in its belief that more radical solutions were called for, which means more democracy and swifter measures to move toward a market-driven economy.[4]

Freedom and turmoil in Eastern Europe will not necessarily be mirrored in the Soviet Union, though it is hard to see how the disintegration of the Soviet internal empire is to be halted. The Soviet Union is hardly likely to reestablish its imperial sway in Eastern Europe without risking a global conflict and the end of any hope of improving economic ties to the West. What is certain is that Gorbachev will inherit the historical responsibility for inadvertently making it possible for the nations of the Soviet bloc to seek their own destiny. At this juncture in history, it may well be that the best thing the United States can do for Eastern Europe, as Czechoslovakia's president, Vaclav Havel, said in an address to the U.S. Congress, is to help "the Soviet Union on its irreversible but immensely complicated road to democracy."[5]

The effect of a history of repression on the peoples of Central Europe was, in a sense, to put history on hold—at least as we in the West know it. For even during the interwar period, with the singular exception of Czechoslovakia, the peoples of Central Europe did not develop along democratic lines. The nationalities, which were loosely held together by the sometimes cruel, generally benevolent, but certainly hollow Austro-Hungarian and Ottoman empires, eventually gave way after 1918 to the dictatorships of General Pilsudski in Poland and Admiral Horthy in Hungary, and the rump Austrian state that was easily annexed by Adolf Hitler in the 1930s; only democratic Czechoslovakia, even though beset with a sizable German minority, seemed to fulfill the Wilsonian dream of self-determination for all peoples. The dominant themes of the Central European experience in the modern era have been foreign occupation and oppression; controlled economies, whether mercantile or Marxist, and in either case routinely unsuccessful; romantic nineteenth-century nationalism; an interminable struggle among ethnic solidarities; and deeply politicized Catholicism and Protestanism, often accompanied by vicious anti-Semitism.

Central Europe, then. Not the Balkans. Here, the history of liberal democracy is, if anything, even more discouraging. Rumania, controlled by a landed aristocracy, gave way to a personal dictatorship by King Carol on the eve of World War II and became allied with Nazi Germany. Even when it was controlled by the Communist party after the war, Rumania was ruled over by the megalomaniacal Ceaucescus—"socialism in one family"—that reduced Bucharest, with its ersatz Parisian charms, to a grim twilight city where food and electricity were severely rationed. The 1989 upheaval that toppled the Ceaucescu family, while sparked by a popular uprising, resembled a military coup more than a broadly based revolution.

The National Liberation Front, which took over as the ruling body of the country and was later elected to govern, was composed mainly of "reform" communists, hardly the type of leadership that would be willing to make drastic economic changes along the lines of Poland or Hungary. The 23 million Rumanians led wretched lives during the last decade of the Ceausescu tyranny: in a richly fertile land, food was rationed, cities were without streetlights, hospitals lacked medicines, and public transportation was a disaster. In order to be self-sufficient (which also meant independent from Moscow), the Ceausescu regime borrowed heavily from abroad to invest in steel mills, petroleum refineries, and other heavy industries. Then, after driving his country into near bankruptcy, Ceausescu decided to pay off the $100 billion debt entirely—and did so. As a result, the Rumanians start off in their brave new world without the burdens of their neighbors. But where they are heading is far from clear, and far from encouraging. The old guard has remained in power after free elections and has used its power to suppress open dissent.

Prospects for stability in Rumania are also threatened by the presence of an oppressed, sizable, and restive Hungarian minority in Transylvania. In addition, the multiplicity of political parties and groupings, combined with the heritage of fascism from the interwar period, bodes ill for the transition to liberal democracy. Yet the desire not to return to the communist totalitarianism of the past is widespread and deeply felt.

Bulgaria, on the other hand, has traditionally remained tied to Moscow through its religious and cultural history, seeing Russia as its protector against the oppressions of the Ottomans. Even after the "reform" communists rid themselves of their Stalinist heritage, there has been only a hesitant movement toward a market-driven economy. In addition, efforts to recognize the rights of the Turkish minority have led to a backlash among ethnic Bulgarians.

Yugoslavia, hardly possessed by any national consciousness, has been an even more profoundly divided country than Rumania or Bulgaria. Among the six republics that were stitched together at Versailles, there are grave economic disparities— from the rich republic of Slovenia in the far north (with annual per capita income of $5,700) to the wretchedly poor Kosovo in the south (where the average is $750). By 1990, democratically elected, non-communist governments held power in both Slovenia and Croatia. A year later, Slovenia and Croatia first declared their independence from the fragile Yugoslav nation, then bargained for a looser confederation of sovereign states, all the while resisting a military takeover by the Serbian-dominated Yugoslav army. The European Community feared that Yugoslav separatism could well spill over to unsettle the minorities in some of its own countries, as well as produce an influx of refugees streaming across the borders into Italy and Austria.

While it is hardly likely that internal Yugoslav violence, or even civil war, would ignite a larger conflagration in Europe, the European governments were aware that they should do everything they could to persuade the Yugoslavs to sort out their differences peacefully—or let the dissolution of Yugoslavia proceed in some order if no other way could be found to resolve Yugoslav tensions. Yet with the exception of Slovenia, all the republics have sizable minorities, and if Yugoslavia were to fall apart, even if this were accomplished peacefully, it would more likely degenerate in the not-too-distant future into a latter-day Lebanon rather than a group of autonomous, nationalistic states.

If it is difficult to find a common culture in the Balkans, what do we mean, beyond the geographical entity, by Central Europe? By repudiating communism, does this mean there is no common heritage, no shared ideas? Are the Central Europeans merely a collection of disparate peoples seemingly yearning for freedom? Vaclav Havel has written of "a certain distinctive Central European skepticism," which he claims "is inescapably a part of the spiritual, cultural, and intellectual phenomenon that is Central Europe."[6] He refers to "a Central European mind, skeptical, sober, anti-utopian, understated." Perhaps this is so now. But if so, it is the product of the years since 1948 when the Central Europeans came under the control of the Soviet Union. Hitler, too, came out of Central Europe, as did the pogroms of the Polish state, and Bismarck's Prussia, which went on to unify Germany through blood and iron.

Yet there is a set of attitudes peculiar to Central European intellectuals that comes out of Central European history—what the British writer Timothy Garten Ash calls "the experience of small nations subjected to large empires, the associated tradition of civic commitment from the 'intelligensia,' their habit of irony that comes from living in defeat—but above all it has to do with their own direct, common, and unique experience of living under Soviet-type systems since Yalta." In this respect, Central Europe is not a region whose boundaries can easily be traced on the map. It is "a kingdom of the spirit."[7]

For the West Europeans, who trumpet their goal of full economic union by 1992, this other Europe is both an opportunity and a vexing problem. Cut off from their common European home, to use Mikhail Gorbachev's phrase, the Central Europeans want to reestablish their lost connections. They want their rich cousins in the West to buy their goods and their food; they want loans and credits, francs and marks and guilders and lire. But to join the Economic Community will not be easy. The economic straitjacket of organized planning has to be fully discarded if the Central European businessmen want to operate in Western Europe after 1992, when capital will move freely, where banks, licensed in their home countries, will be able to open branches in any city from Edinburgh to Athens. In short, while Western Europe is out to construct a world with new rules, Central Europe has to undertake the Herculean task of dismantling forty years of Kafka-esque regulations.

And what of the ideology of equality—socialism's great ideal that is part of the habits of mind of the Central Europeans? Can the Polish workers of Solidarity temper their upbringing and become good social democrats whose policies hardly differ from the moderates of West Germany and France? Lost connections take time to be retied, and in the meantime there are the atavistic nationalisms that rage among the Hungarian minorities who have been harshly repressed in Rumania, the Germans who were given over to Poland after World War II, and everywhere the presence of the former agents of Soviet imperialism, collaborators in repression.

The greatest danger over the next few years for Eastern Europe stems from the consequences of economic failure. Putting aside East Germany, which has become part of a greater Germany, the nations of Eastern Europe, from Poland to Bulgaria, are all determined to move toward market-driven economies. The Poles have taken the boldest step by invoking a form of "shock therapy." On January 1, 1990, by freeing prices and doing away with subsidies while continuing wage controls, the Poles chose to subject their economy to a virtual overnight transformation into a market economy. The initial pain was great: inflation soared to 70 percent in one month; with purchasing power dropping and the government keeping a tight rein on the money supply through tight credit and a balanced budget, sales slumped and bankruptcies and payoffs resulted. Poland's economic output fell by 13 percent in the year after its radical economic program took effect.

Despite the pain, there were also signs of long-term gain. Firms cut back prices as they learned the way the marketplace operated and sought to attract customers anew. Absenteeism dropped by half. Product shortages that had been characteristic of the communist economy started to disappear. Moreover, despite widespread strikes, the Poles pushed on. In July 1991, the government took another giant step toward privatizing the economy by announcing plans to transfer a fourth of its state-owned industry into private hands. This would be done by handing over 60 percent of the equity in each of the 400 state-owned factories to about twenty investment funds. Ownership of the funds, in turn, would be given to every citizen over eighteen years old. The investment funds would be run by financial experts from the West rather like Western-style mutual funds.[8]

If the Poles can maintain the discipline to withstand the changes that are taking place, if foreign investment and credits flow in, then the government may be able to hold together without alienating its own support at the grass roots. But as a result of high unemployment and growing disparties between the newly rich and the newly poor, there are likely to be continued, severe challenges to the government. Yet President Lech Walesa campaigned on the need for the govenment to move faster in its program of privatization and marketization. He also threatened to ask for powers to rule by decree and for the removal of communists from the bureaucracy. In the transition to democracy, the tensions within Poland point up the dangers ahead throughout Central Europe: the economic changes under way will prove deeply painful, and the temptation to authoritarianism is always present.

Like Poland, Hungary has been moving toward a free economy. Moreover, the Hungarians had already made some progress along these lines through the cautious reforms instituted by party chief Janos Kadar after the 1956 Hungarian Revolt was crushed by Soviet tanks. But his partial decentralization of the economy—so-called goulash communism—turned into a Rube Goldberg apparatus that never quite worked, even though, relative to other members of the Soviet bloc, Hungary's economy prospered. While the private sector eventually accounted for almost 25 percent of Hungary's economic activity, there was still too much centralization. By the mid-1980s, Hungary had broken down under a mountain of debt and rising consumer shortages; in 1989 it had a near 20 percent inflation rate, rising unemployment, Europe's highest per capita foreign debt, and a falling standard of living. Nonetheless, the Hungarians became committed to privatization and opened the first stock exchange in Central Europe.

But in Hungary, unlike in Poland which had a broad-gauged commitment to the new economics, divisions over how to achieve a market economy sprang up within the anticommunist opposition. The 1990 elections may have eliminated the Communist party as a political force, but it left the country to be governed by the Democratic Forum, a center-right party with a strong nationalist, and even anti-Semitic, tinge. Moreover, the Democratic Forum wanted to move more slowly in abandoning socialism than its rival party, the Alliance of Free Democrats.

Despite these tensions, Hungary, like Poland, decided there was no turning back. Its preferred method of proceeding with privatization was to sell off firms to the highest bidder, foreign or domestic. Perhaps for this reason, it remained the favorite among Western investors. By 1991, it was showing a surplus in its current accounts balance while keeping to its schedule of paying off its $21 billion debt.[9]

But if Hungary's economic situation does not continue to improve over the next few years, it is quite possible that a nationalist demagogue will distract the people from their problems by raising the issue of the sizable Hungarian minority in Transylvania that has been badly treated by the Rumanians ever since they acquired the region at the end of World War I.

The main problem for Hungary is a deep lack of tolerance or civility in politics. Some form of reconciliation will be required between the two parties—the Democratic Forum, nationalist and populist, and the Alliance of Free Democrats, more liberal and Western-oriented—if Hungary is to be governed effectively. But if such a reconciliation endures, it will be because it stems from political necessity rather than from any disposition to compromise. Hungary's divisive politics render the likelihood of instability greater precisely because there is no commanding political leader, such as Czechsolovakia's Vaclav Havel, who can bridge the gaps on controversial issues.

Czechoslovakia, whose "velvet revolution" toppled the communist rulers in *four days*, is especially blessed by having the playwright-dissident Vaclav Havel as president. Who else could have traveled to Warsaw and told the Polish parliament that "the most dangerous enemy today is no longer the dark forces of totalitarianism, the various hostile and plotting mafias, but our own bad qualities"? Czechoslovakia, the only wholly democratic country in Central Europe between the wars, one of the most productive countries in Europe in 1938, is the one country in Central Europe most likely to ensure political freedom, though it may not hold together as one country. The circumstances confronting Havel are very different from those facing Thomas Masaryk, the first president, in 1918. When Masaryk came to power his main tasks were to reduce the power of the Roman Catholic Church and the influence of Germany. Havel, on the other hand, is supported by the Church, and Germany is Czechoslovakia's most likely source of economic support.[10]

In one sense, Czechoslovakia, precisely because it had not been allowed to muddle through with reform communism, is a *tabula rasa* upon which a market economy can be built. Let us not forget that there is a memory of entrepreneurship in Czechoslovakia that has not been lost; in places like Russia, Poland and Rumania, with their heritage of landowning aristocracies, such a system hardly existed. So in Czechoslovakia, as in Hungary and Poland, though more slowly, a plan to privatize has been set in motion whereby the government will use vouchers to sell firms directly to its citizens. And by 1991, Western investment was starting to come in, bringing with it new technology, new management skills, and new jobs.[11]

Moreover, the prospects for economic success may vary greatly among the former Soviet satellites. Were Czechoslovakia and Hungary to move more rapidly to a higher rate of productivity and a significantly better standard of living than Poland, to say nothing of Bulgaria and Rumania, the likelihood is for greater tension within the region and the consequent search for scapegoats, which could mean a new outbreak of anti-Semitism and harsher treatment of ethnic minorities. The probability of turmoil leading to interstate conflict is the plausible legacy of a revolution betrayed.

Here, then, is Eastern Europe, miraculously and suddenly freed from the domination of the Soviet Union, but also freed from the apparent stability that the two superpowers imposed on a divided Europe. The danger is that chaos will prevail in a multipolar Europe. The U.S. Deputy Secretary of State Lawrence Eagleburger seemed to believe this likely when he said that such a world is not "necessarily going to be a safer place than the Cold War era from which we are emerging." He went on to warn: "For all its risks and uncertainties, the Cold War was characterized by a remarkably stable and predictable set of relations among the great powers."[12]

His Cold War nostalgia aside, Eagleburger's pessimism is not wholly misplaced. His is an almost Hobbesian view that without a dominant power to keep order, disorder arising from political anarchy is likely to prevail. The ethnic tensions in Eastern Europe, the tradition of authoritarian rule, the ravages of the economic system imposed by the Soviet Union—all these elements might well bring forth new forms of authoritarianism rather than compromise, harmony, and liberal democracy. Unfortunately, Eastern Europe cannot simply be put into quarantine if these patterns become dominant. Aside from the threats to Western economic investment, there are the legitimate security concerns of the Soviet Union and the domestic lobbies in Western Europe that will not allow governments to abandon their spiritual cousins in the East.

To counter this pessimistic outlook, there are the examples of the civil societies of Western Europe, the memory of coercive communism, and the desire felt by virtually all East Europeans to belong to a larger European polity. By this they mean the European Community. A stable political order in Europe leading to civic democracy requires tying Eastern Europe to the democratic institutions of the European Community. Just as Spain and Portugal were permitted to join the Community as long as they maintained liberal democratic regimes, so, too, the transition to democracy in Eastern Europe will best be served if the East Europeans are first accorded associate membership.

For the United States, a stable Europe requires a residual

American military and, above all, an American economic presence in Europe. Once again, a solvent America, with ample financial as well as military resources, is the only way the United States can play a central role in the transition to democracy of the former Soviet satellites. It also means that the United States should seek the participation of the Soviet Union in the security of Europe as part of a pan-European security organization, and, insofar as Soviet economic trends permit, allow the Soviet Union to belong to international economic institutions.

The most successful example of how international institutions helped to promote liberal democracy was, after all, the Marshall Plan, precisely because it was accompanied by a host of new economic organizations. These new institutions included the European Coal and Steel Community, the General Agreement on Tariffs and Trade (GATT), and the European Payments Union. The United States did not just buy stability and liberal democracy; the Marshall Plan worked because it created international organizations that helped to channel domestic interests in a direction favorable to international cooperation and stability.[13]

In post-Cold War Eastern Europe, Western loans and investment by themselves are inadequate; but the new European Bank for Reconstruction and Development that will make loans to Eastern Europe to aid in the creation of market-oriented economies is the right move at the right time. The bank is to have an initial capital of $12 billion, with forty-two countries of the East and West as founder members. The United States will be the largest shareholder, with 10 percent; Japan will follow with 8.5 percent; and the European Community will control 51 percent of the capital.

For East Europeans, the path to prosperity lies in association with their counterparts in the West. The French under President François Mitterrand have been convinced that pushing European unity forward at high speed will prove the best way of anchoring Central Europe to the West. Before concluding association agreements with Poland, Hungary, and Czechoslovakia, the French—and other West Europeans as well—believe they must take further steps toward European unity, the so-called deepening of the Community. They insist that in the long term they can satisfy the yearnings of the Central Europeans to become part of the European Community while pursuing their own agenda of political union.

But there is a potential obstacle to the construction of a tightly organized Europe. Not surprisingly, its name is Germany. For unlike the other nations of Central Europe, East Germany became part of the European Community simply by uniting with the Federal Republic. German Chancellor Helmut Kohl has tried to reassure the Community that "this German union will be seen as a push and not as a brake for the Community." He said in Brussels in 1989, "We don't want to be the Fourth Reich. We want to be European Germans and German Europeans."[14] Other Europeans and Americans have not necessarily shared Kohl's beliefs. In France, a leading businessman, Alain Minc, thinks that with a reunited Germany inside the Common Market "the process of building Europe belongs to the past."[15]

On the other hand, it was precisely the need to digest Germany that forced the German chancellor and the French president to speed up progress toward European union. Comprehensive new agreements on economic, monetary, and political union were to be concluded by January 1, 1993. That will probably not happen according to the timetable that the Brussels bureaucrats would prefer—not least because the French are reluctant to admit Central Europe to the Community, which they see as enhancing German power. But by the end of the century, the Community will be stronger than ever. It will also be a German-dominated Europe. As the two Germanies became one, most statistics about a greater Germany simply lumped together the existing populations and productive capacities of the two parts. This is a formidable proposition in itself. Under these calculations, a united Germany would total 78 million hard-working inhabitants. It would export $354.1 billion (compared with America's $321.6 billion); its balance of trade would produce a surplus of $73.9 billion (compared with a projected U.S. trade deficit for 1990 of just under $100 billion); and by 1995, its total gross domestic product would exceed that of France and Britain combined.[16]

But this is only the beginning. We have to imagine a far more productive East Germany, freed of the shackles of centralized planning and limited incentive. East Germany was already a heavily industrialized economy before unification, and East and West German machines were built to the same specifications and are compatible. A fully productive Germany will mean a German-dominated Europe—to be sure, not the Bismarckian or the Hitlerian reichs, but nonetheless the Europe of greater Germany, whose economic policies will be organized by the Deutsche Bank and the Bundesbank, and whose monies will flow into Central Europe, as they did just before World War I. But even a democratic Germany, whose capital and goods rather than its soldiers and weapons are the yardsticks of its power, is a Germany that will set the political agenda for the new Europe, which will by its very nature profoundly affect America's national interest.

PART II

The New European Order

4

The German Question

In delineating the contours of the new Europe, which will now require the integration of the newly emancipated nations of what was once Soviet-controlled Europe, the economic and political weight of Germany inevitably dominates the landscape. As John Maynard Keynes wrote of Germany in 1914: "Round Germany as a central support the rest of the European economic system grouped itself, and on the prosperity and enterprise of Germany, the prosperity of the rest of the Continent mainly depended. . . . The whole of Europe east of the Rhine thus fell into the German industrial orbit, and its economic life was adjusted accordingly."[1] Today, Germany's economic ties to Central Europe are closer than any other Western power's. The German question is not who will dominate Central Europe, for surely Germany will do so, but whether the new Europe can accommodate a unified Germany. In short, is Germany a satisfied power?

It was German might, after all, that destroyed the preeminence of Europe and surrendered the old continent to the new leviathans—the United States and the Soviet Union. It is Germany that must be contained—without humiliation but also without hypocrisy—by the Europeans, the Russians, and the Americans, just as the Soviet Union must be tightly bound to Europe in a new quest for collective security. The containment of Germany, however, should not be viewed as a response to a thrusting expansionist German drive for military power. Germany at the end of the twentieth century does not resemble Germany at the end of the last century. It is not a Wilhelmine Germany devoted to the continuous assertion that Germany is powerful and great.[2] Nor is it a Hitlerian Germany, whose aims were to overturn the *diktat* of the Versailles Treaty and to assert military control over all Europe. Yet no European—and, one hopes, no American—political leader should rest content without a policy that anchors Germany firmly in the West: militarily to the United States as well as to any all-European defense grouping, and economically and politically to the European Community, in which a strong France holds a central place.

The prospects for a democratic, pacific Germany are more promising than they have been since the German Reich was created by Otto von Bismarck over 120 years ago. Fortunately, Germany has had two great statesmen since that time. Bismarck was one, but surely Konrad Adenauer was the other. Adenauer could not have been more different from Bismarck, not least because he was a Catholic and a Rhinelander who disliked the domination of Bismarck's Prussia. He has been described as wanting "in the early 1920s what he achieved after 1945—a Rhineland separated from Prussian rule, remaining within the Reich, but with a special economic relationship with France."[3]

Adenauer's great objective was to integrate the Federal Republic with Western Europe and to gain for it a status equal to Britain and France. Fearing the domination of the Soviet Union, he also wanted to bind West Germany to a military alliance with the United States. By anchoring Germany in the West, he seemed to think that a prosperous Federal Republic would prove a magnet to the centralized economies of East Germany and Central Europe. Whether he actually believed that reunification would only take place through a "policy of strength" by the Western allies, coupled with West German prosperity, is impossible to know. This is what he said, and the erection of the Berlin Wall in 1961 seemed to signal the failure of his policy, though it really was a testimony to its success with the East German populace.

Despite his apparent belief that reunification would only come through a strengthened Western alliance, he did have reunification plans secretly drafted. In 1958, he proposed an "Austrian solution" that would call for a reunified neutral Germany, and in 1960, he suggested separate referenda in both German states on reunification. Not surprisingly, Nikita Khrushchev rejected both ideas. What Khrushchev wanted was a weaker West, and Adenauer's proposal would have led to a stronger system in the West.[4] By the 1960s, a reunified Germany ceased to be a possibility—if indeed it ever was. The kind of Germany that might have led to reunification—a neutral, disarmed Germany—had simply disappeared by Adenauer's death in 1967.

The policy of an opening to the East—*Ostpolitik*—succeeded Adenauer's emphasis on arming the Federal Republic within the Atlantic Alliance. With the signing in the early 1970s of treaties with the Soviet Union and its Warsaw Pact allies that regulated Bonn's relations with the U.S.S.R., the Federal Republic accepted the postwar status quo. This meant that relations between West Germany and the East European states could be normalized, including relations with East Germany. A new agreement over Berlin turned out to be generally favorable to Bonn, because it made access to and from Berlin much easier. In short, by the end of the 1970s the way was opened for an increasing West German political and economic involvement in East–Central Europe.

Because of its German connection, East Germany already had special arrangements that allowed it access to the Common Market; in addition, Bonn had already extended it an annual subsidy of between DM 4.5 billion and DM 5 billion. The reality was that East Germany *was* a captive nation—of West Germany. And West Germany, now the fourth greatest economic power in the world, ranking after the United States, the Soviet Union, and Japan, had become the West's biggest trading partner with Eastern Europe; at least 50 percent of East European trade with the West was with West Germany. By 1988, West Germany had also become the paymaster of Eastern Europe, as it guaranteed a DM 1 billion credit to Hungary, which helped to keep Hungary from defaulting on its Western debts.[5]

By the end of the century, the new, reunified Germany is likely to be an economic power second only to the United

States. With the continuing exposure of economic weakness in the Soviet Union, only Japan shows signs of maintaining its lead over Germany in overall gross national product. But even when East Germany was hobbled by Stalinist planning and centralization, the East Germans became the most productive country in Eastern Europe. The notion that the East Germans are somehow infected with communist lethargy neglects the calculation of national character. "We are not 61 million people in central Africa," a minister of the West German government told the American commentator Sidney Blumenthal in 1990, "we are 61 million Germans in Central Europe."[6] The new Germany, let us not forget, will number some 78 million. The merging of the German economies, including the monetary union that replaced the East German currency with strong West German marks to produce a common currency, only marginally strained the West German economy; it did not set off world inflation as some feared. For this reason, Bonn forestalled economic and social chaos and paved the way for construction of the German locomotive that will drive both Eastern and Western Europe.

A unified Germany is likely to be a satisfied power. The aim of Western policy, however, is to make sure that it remains so. So, too, is this the central goal of Soviet policy, and would also likely remain so under any Soviet leader. "A united Germany," as Gorbachev put it, should not "spell a threat or harm to the national interests of neighbors, or anybody else for that matter."[7] With East Germany expected to become, at least in the short run, a virtual West German colony, the likelihood is that Germany's culture will be become more assertive. Germany's presumed economic and cultural hegemony is already recognized, as Czechoslovakia's ambassador to the United States, Rita Klimova, acknowledged. "The German-speaking world"—by which she meant a unified Germany plus Austria—"will now achieve what the Hapsburgs, Bismarck and Hitler failed to achieve: The Germanization of Central Europe."[8]

The moral authority of the European Community is therefore as necessary for Germany as it is for the smaller powers of Central Europe and the Balkans. Both Chancellor Kohl and French President Mitterrand have repeatedly stressed the need to tie Germany to the West. Kohl has time and again reiterated his belief that "there is no contradiction between German unity and European integration."⁹ At the same time, like their Russian counterparts, Kohl and Mitterrand are eager to avoid creating any paranoia in Germany by constantly reminding the Germans that the West and the East are fearful of Germany's economic, political, and military potential.[10]

An anchoring in the West means not only acceptance by Germany of the Council of Ministers in Brussels, the Parliament in Strasbourg, and the Court of Justice in Luxembourg, it also means wrapping Germany not simply within NATO but within a broad security organization that would include both U.S. and Soviet armed forces (which I discuss at greater length in Chapter 6). This may well be Germany's own preferred solution.[11] An arrangement for collective security that would include the Soviet Union will allay Soviet and European fears of both German revanchism and German military and economic power. It would facilitate what today's German leaders see as their mission: keeping the Soviets in Europe, not as an army of occupation but as a political, economic, and cultural presence.[12]

For it is not the neutralization of Germany that would satisfy the legitimate security concerns of both East and West, but the maintenance of Germany within an alliance system that would circumscribe German ambitions and, above all, the level of German militarization.[13] The Germans, in turn, are aware of their own need to receive guarantees of security. Only a European security organization, backed up by an American nuclear guarantee, would fulfill these requirements.

Moreover, the Germans realize that Russian fears of a unified Germany have to be assuaged. The Kohl government has already offered to be the Soviet Union's chief banker and benefactor. It is committed to providing DM 70 billion (about $40 billion) to the Soviet Union over the next few years; this is comparable to the U.S. commitment under the Marshall Plan. It will pay hundreds of millions of dollars to get the 380,000 Soviet troops out of East Germany over the next three to five years. It has agreed to carry out the terms of all East Germany's trade accords and fulfill all contracts that East German firms had with the Soviets.

In addition, Bonn said it would limit the size of the armed forces of a united Gemany, agreeing to restrict the future German military to 370,000 troops, compared with about 667,000 in 1990 for the two German armies combined. Finally, the West

Germans said that no nuclear weapons would be deployed on what has been East German soil during the transition period, and probably not afterward.[14]

Does this mean that the drift of German history will lead to a new "Rapallo," a German–Russian deal that will turn the Germans away from the West? Certainly the importance of Germany's links to the West will diminish as Germany's ties to the East grow stronger, especially after the transfer of Germany's capital to the city of Berlin.[15] But history does not repeat itself. Like generals who prepare to fight the last war, we too often expect a similar peace. To many of us, a unified Germany seems a disruptive weight in the European balance of power. Yet for half a century after Waterloo, the nations of Europe feared a resurgent, aggressive France, a France that had ravaged Europe for almost 200 years. But after 1815, France never again tried to exert military predominance over all Europe. When Germany finally achieves all the economic, social, and even military ends of unification, it may well become at long last a sated power, content to manage the economies of Europe, while a wary Russia and America ensure that any warlike German ambitions are firmly held in check.

5

The European Challenge

"We see a foreign challenger breaking down the political and psychological framework of our societies. We are witnessing the prelude to our own historical bankruptcy. At times like this we naturally think about reinforcing the barricades to hold back the invader. But purely defensive measures might well make us even weaker. In trying to understand why this is so, we stumble across the key element. This war—and it is a war—is being fought not with dollars, or oil, or steel, or even with modern machines. It is being fought with creative imagination and organizational talent."

These words come from the biggest bestseller in France in the 1960s—*Le Défi Américain* (*The American Challenge*) by Jean-Jacques Servan-Schreiber. Europe, Servan-Schreiber believed in 1968, was in decline. Unless the European Common Market moved toward integration, Western Europe was doomed. As American historian Arthur Schlesinger, Jr., put it at the time, "the present European generation has only a few years to decide between restoring an autonomous European civilization or allowing Europe to become a subsidiary of the United States."[1]

Could any Frenchman write these words today? It seems highly improbable. Most likely, the author would be an American, fearful of Europe's economic strength and worried about America's solvency, burdened as America is with the twin towers of the fiscal and trade deficits. Europe sees an America struggling to define a new role for herself in a Europe that is planning an economic take-off as it expands to encompass the remnants of the Soviet bloc and the neutral nations clamoring for entrance. Predictably, it was the German chancellor who said at the 1990 Harvard commencement that the new European federation "should not be an exclusive club confined to the present members of the European Community," but "open to Poles, Czechs, Slovaks, or Hungarians" as well as "to countries like Austria, Sweden, Norway and Finland."[2]

When Servan-Schreiber wrote *The American Challenge,* he was bewailing the decline of Europe before seemingly inexorable U.S. economic and political might. Europe, however, took up his thesis and turned it on its head. The twelve nations of the European Economic Community are now prepared to move toward a single market. Moreover, they expect to ratify agreements on political, monetary, and economic union in the 1990s, to coincide with the completion of the internal market. In so doing, the European Community now looms as a superpower of staggering economic productivity, its 344.6 million people (including a unified Germany) capable of producing an annual output of goods and services of $5.53 trillion, exceeding even America's ($5.47 trillion) in 1990.[3] With the added growth and productivity provided by the single market and the new markets of what used to be called Eastern Europe and countries of the European Free Trade Association (Sweden, Switzerland, Norway, and Austria), Europe's economic weight might soon carry it to the head of the world table. It would mean a total European population of about twice that of the United States and four times that of Japan and the four Asian "tigers"—South Korea, Singapore, Hong Kong, and Taiwan. The emerging European bloc will likely account for approximately one-fourth of the global product.[4]

In addition, a governing body evolving within this Community will likely produce a common central bank before the end of the decade, some form of common currency, and a fully unified trade policy. From this image arises the even more frightening specter—to Americans at least—of a Fortress Europe dominated by great industrial groups that could freeze all competitors out of its market. If this should happen, the risks to the United States would be huge, for American producers sell almost one-fourth of all their products to European consumers.

At the moment, the United States and Europe operate along a two-way street, or to use a more appropriate nautical term, understand how to navigate among the shoals of self-interest. In 1988, trade between America and the European Community came to more than $160 billion, and combined direct investment in each other's market was estimated at some $329 billion. The United States, however, has been gravely dependent on borrowing European capital to balance its books. In 1988-89, for example, a net $153 billion in capital from the European Community flowed into the United States as direct investment in plant and equipment and as bank deposits and cash to buy government notes and corporate stocks and bonds. (Japan, by comparison, provided $104 billion over the same period.)[5]

Despite this Keynesian picture of free markets leading to global prosperity, protectionist pressures have been growing as the Europeans go on constructing mini-forts to provide selective protection against the threatening dragons of Japan and East Asia, or even against the United States. Cars, for example, remain the great test case. Five countries set up quotas limiting the import of Japanese cars to a very small proportion of their car market: France, Italy, Britain, Spain, and Portugal. The vice-president of the European Community Commission said that national restrictions on car imports would be abolished by the end of 1992, but he also admitted that his words were "more than a hope and less than a decision."[6] In 1989, a skirmish broke out when the Community imposed a ban on $130 million worth of beef from hormone-treated cattle. Then in 1990, the Uruguay Round of international trade talks was suspended, largely because Germany and France refused to lower trade barriers on agricultural imports, a policy that especially hurt Third World food-exporting nations.[7]

Not only in Europe but also in North America and the Far East, the emergence of three trading superblocs is in the making. In 1988, Washington signed the historic free-trade agreement with Canada, and, if the U.S. Congress permits, a free-trading zone will be extended to Mexico. Since Canada is the largest trading partner of the United States and Mexico the third largest, a North American bloc is virtually in place. In the Far East, Japan already evokes chilling memories of World War II's Greater East Asia Co-Prosperity Sphere as Tokyo forges closer ties with the newly industrialized economies of Singapore, Taiwan, Hong Kong, and South Korea.

The grave social problems the United States is experiencing do not bode well for competition with the new Europe, even should the Community keep its barriers down, as most European industrialists and Brussels-based Eurocrats desire. America's infrastructure is crumbling. (Every two days a bridge collapses somewhere in the United States.) About 20 percent of America's eighteen-year-olds are now functionally illiterate, which hardly puts the United States in a good position to compete with a better-educated Europe. U.S. government spending on commercial research and development has gone down almost 95 percent over the past twenty years. Rates of savings and investment as a percentage of GNP are among the lowest in the industrialized world.[8]

Moreover, as the extent of the U.S. military commitment to Europe diminishes in the face of Soviet withdrawal, so too will the ability of the United States to have its way with Europe, were Europe to pursue a hostile policy toward U.S. economic interests. Our ability to control events in Europe for the past half-century has been directly related to the Europeans' belief that they needed our protection from Soviet aggression and intimidation. This made the Atlantic Alliance possible.

As European economic power increases and political cohesion follows in its wake, the ability and desire of the Europeans to provide for their own security are also likely to grow. This will certainly facilitate a further lowering of the U.S. military presence in Europe. Over the next few years, U.S. troop strength on the continent will drop to perhaps as low as 50,000 troops, as German unification is accompanied by severe cuts in Warsaw Pact and NATO armed forces as part of conventional force agreements with Moscow. Moreover, the war in the Persian Gulf, when America took the lead in organizing and directing a military coalition, may well convince most Commmunity governments that they must speed up their approach to integration if Europe is to emerge as a major economic bloc that can also play a major politcal role in the world. In this instance, Europe still had to rely on the United States to defend its vital interests; this was an embarrassing moment for most Europeans, but as a result, as the French foreign minister noted, "it seems that we now have better prospects for political and security union."[9]

For the committed Europeanists, the likelihood that Europe will emerge as a superpower in its own right (with its own arsenal of nuclear weapons) will depend on how successful the Common Market is in deepening itself along the lines Mitterrand, Kohl, and Delors had once sketched out. Some economists have predicted that the benefits of 1992 may be as much as five times greater than the European Commission anticipated in 1988, when the Cecchini report anticipated an increase in the European gross domestic product of between 2.5 and 6.5 percent. Three great events—the creation of the single European market, the unification of Germany, and the opening up and rebuilding of Eastern Europe—could easily push up the rate of European growth to an average of 3 percent or more for the rest of the decade.[10]

By removing the right of a single country to veto progress toward a single market, the European Commissioners believe that their agenda will commit the nation-states of the Common Market "to transform relations as a whole among their States into a European Union." Delors has already said that by the year 2000 "80 percent of economic, and perhaps the social and tax, legislation will be of Community origin." He also predicted that an all-European government would be in place.[11]

There is, however, a central question at the heart of this scenario. All of these projections were composed before the fall of the Soviet empire in Eastern Europe. The old image of Europe 1992, powerful but possibly perfidious, is abruptly out of date. Now the European Community will have to accommodate the new democracies of Central Europe and swallow the humongous new German state. Yet the Brussels bureaucrats maintain that the Community can "deepen" and "widen" at the same time. And they may well be right. The French policy of tying Germany into a European setting, far from being challenged by the Germans, is seen by most Germans as realism. At the same time, France may well try to frustrate Germany's desire to grant Central European nations access to the Community. As both Paris and Berlin agree, the goal must be, in the words of Thomas Mann, "a European Germany and not a German Europe."

The challenge the Europeans now face is to devise creative political and economic structures that can ensure that German power and prosperity are absorbed into a broader European power-sharing arrangement. Otherwise the dominance over European economic policies exercised by the Bundesbank will only grow with German unity. Every financial and economic decision made in the months preceding German monetary union in 1990 was made in Bonn, or in consultation with Berlin, not Brussels. As the single market falls into place and eliminates many of the tools previously used by national governments to maintain their independence, German dominance will only increase. The potential nationalist backlash from such a situation could put an end to Europe's resurgence, a threat both Germany and its partners are well aware of. Either way, a united Germany will be the centerpiece of the new Europe, and its investments will continue to flow into Central Europe, as they did before the First World War.

So there will be a European challenge after all, but it need not be a destructive one. It may take its character from Germany, but this can still mean it will be a democratic Europe, one that will make every effort to ensure that all the member states of the Community are democratic as well. Spain, Portugal, and Greece—all functioning democracies—attest to the power of the European ideal. European unity may take more time than the Eurocrats would like; but the Europe that emerges will likely be a stronger one, led by Germany but nonetheless anchored within a greater European polity.

Is there a role for the United States in this new Europe? If there is not, then the political outlook for Europe will almost surely darken. The United States, still a great economic and military power, remains indispensable in ensuring the stability of Europe from the Atlantic to the Carpathians. Where once American forces were needed primarily to deter a Soviet attack from the East, the Americans are now needed to reassure the Germans that they will not find themselves vulnerable to Soviet pressure, and hence tempted to build up their own military forces, even to the extent of equipping themselves with nuclear weapons.[12] The Americans will also be useful for reassuring the other European nations and, for that matter, the Soviet Union itself, that Germany will not again threaten the stability of Europe. An American commitment to the security of Europe, as political scientist Michael Mandelbaum has written, "would be designed not so much to deter an immediate threat from the Soviet Union as to reassure all of Europe—including Germany and the Soviet Union—that it need not fear a power vacuum. Such a vacuum might compel European nations to recalculate their military requirements, perhaps in ways others would consider as threatening."[13] America's goals in the post-Cold War world are clear: to retain significant American influence on the continent; to assist in the liberalization of the Soviet Union; and to make sure that the new Europe will not be hostile to American interests.

Nor can there be an enduring structure of peace without the participation of the Soviet Union. Should the United States and the Europeans try to construct a Europe to exclude the Soviets, then the growing fears on the part of the Soviet Union that it is threatened by economic and political ostracism are likely to bring about the very thing that the Europeans, and especially the Germans, fear most—that once again the Russians will upset the fragile stability of the continent.

The balance of power is not an anachronistic concept in Europe, even as supracapitalistic economic structures centrally affect issues of peace and prosperity. The United States still needs a strong Germany to counterbalance the Soviet Union in the east. We also need a stable Soviet Union to help contain the forces that have been unleashed in Eastern Europe by the disintegration of the Soviet empire. And to ensure this, we need to promote democracy within the Soviet Union.

Above all, for America to retain the respect of Europe and the power to back up that respect, it is vital for the United States to put its own economic house in order, so as to rid the Europeans of the impression of American weakness stemming from economic profligacy. But even while the key components of power in the twenty-first century are likely to be economic and not military, the military presence of the United States abroad, undergirded by a solvent economy, would ensure the balance of power in Europe and therefore the global balance that is already in the making. America's European commitment remains just as valid today as it did yesterday, but the nature and degree of that commitment must be almost wholly reinvented.

6

The Concert of Europe

"I T'S the best I can do for Poland at this time," Franklin Delano Roosevelt protested to his chief of staff at the 1945 Yalta conference. He had obtained Stalin's promise for "free and unfettered elections" in Poland. He could do little more given the presence of the Soviet army in Eastern Europe that February. Yet within a few months of the Yalta accords—what Harry Hopkins, F.D.R.'s closest adviser, had called "the first great victory of the peace"—the United States accused Russia of betraying the spirit of Yalta. In fact, there never were free elections in Poland until 1989. The Cold War in Europe, which had begun in a dispute with the Soviet Union over the disposition of Central Europe, became history only when the peoples of Central Europe were finally able to decide for themselves the legitimacy of their governments.

The states that once composed the Warsaw Pact, allied to Moscow in name only after 1989, became hollow armies without a mission. No one would fight to defend Russia. In addition, the East-West negotiations to obtain a balance of forces from the Atlantic to the Urals, the conclusion of agreements eliminating medium-range nuclear missiles and cutting back on long-range strategic weapons, and the near certainty that battle-field nuclear weapons will be eliminated have reduced the need for large standing armies on the European continent. The question that will, for some time to come, plague the West as well as the East is therefore to define the nature of a European security system.

European security has not been settled simply because a unified Germany has become a member of NATO. Soviet President Mikhail Gorbachev gracefully accepted the inevitable when he declared at a spa in the foothills of the Caucasus Mountains after talks with the West German chancellor on July 16, 1990, "Whether we want it or not, the day will come when the reality will be that united Germany is in NATO."[1]

Gorbachev's words were an admission that the West had won the Cold War. The celebration of that victory, however, should not result in an alliance that will keep the Soviet Union out of Europe. Rather, what is needed is a military organization that will keep the Soviet Union in. Secretary of State James Baker seems to have recognized this when he said in 1990 to his NATO colleagues that "the way to build the peace is to reassure the Central and Eastern Europeans and the Soviets that they will not be left out of the new Europe."[2] The ultimate goal of U.S. foreign policy should be to help construct a European Security Organization (ESO) that would include both the United States and the Soviet Union, because it is in the national interest of the United States to do so.

Nor is this a decision that should be left up to the Europeans. America has stood for half a century as the guarantor of the European balance of power. It need not abandon that role, but it does need to adjust to the changing face of Europe by helping to create a military organization that responds to the security needs of Europe, and hence of the United States. The security of Europe may no longer be threatened by an expanding Soviet Union, but it is threatened by turmoil in the newly independent republics of what was once the U.S.S.R., and in the newly liberated countries east of what was once the "iron curtain." It is threatened by nationalistic quarrels that might undo the 1945 borders by force. It is threatened by the possibility both of German might and of Russian revenge over America's "victory" that may have ended the Cold War but may also shatter the long peace.

Surely it is in America's interest that the internal politics of the Soviet Union move in a democratic direction. If we were to isolate the Russians, still likely to remain the other great military power on the globe, we might find ourselves facing an embittered population, increasingly hostile to European prosperity and only too ready to evoke the specter of German revanchism.

Surely it is in America's interest to preserve a European balance of power by keeping Germany firmly anchored in the West and tying the Soviet Union to this Germany. After 1815, the Russians remained the most powerful military force in Europe even though they lagged behind the West Europeans in economic prowess and political evolution and were ruled by a tsar, Alexander I, whose liberal predilections gave way to mysticism and religious orthodoxy. This could happen again. If the Soviet Union is to be a troublemaker, better it should make trouble within an all-European military organization than outside it, and better it should be committed to maintaining the peace in Europe than threatening that peace. While the Soviet Union is likely to be eventually admitted to international economic organizations, its political evolution along democratic lines would be greatly aided if the U.S.S.R. were also embedded in the apparatus of Western security.

Unless the Soviet Union is given an honored place in the common European house, it will be almost impossible for the United States to drastically reduce the scale of its armaments, dismantle its own national security state, and divert its resources to solving those internal problems that threaten to turn America, the inventor of the postwar European political order, into a hollow power with military prowess and economic weakness. We would soon be seen as a political Gulliver bound down by our domestic woes, and more able to service the world with our music videos and our McDonald's restaurants than to enlighten it with our broad culture, our science, and our technology, which have most helped to define the American century.

In the beginning of the Cold War, the United States and the European powers sought to create a European Defense Community that would tame German nationalism, contain Soviet aggression, and further contribute to the unification of Europe. In the 1950s, as in the 1990s, the German question loomed especially large. If the allies were to allow the Germans (still disarmed and under allied occupation) to rearm, it was thought best that this be done by creating an integrated European army combining units from among the Western allies, including the West Germans. This is what the French government believed in the early 1950s, and it soon became U.S. policy. Proponents of European unification expected that a European Defense Community, allied to the United States, would be the military equivalent of the Common Market, both of which would lead to the European Community of 1992, with its single market and its commitment to political unification.

The European Defense Community was stillborn because the British refused to join it (preferring to preserve their illusory "special relationship" to the United States and their ties to the Commonwealth). Without Britain, the French, fearful of German domination, dropped their support in August 1954. As a result, West Germany became a member of NATO, with its own national army—the very thing the French and British had orginally wanted to prevent—in exchange for Bonn's promise not to produce nuclear weapons. In turn, the United States and Great Britain committed themselves to keeping troops on the Continent.

A half century later, NATO endures, but without the mission for which it was created, namely to deter possible Soviet aggression. And while the Europeans may someday revive the idea of a European Defense Community, they are unlikely to press for an American troop withdrawal from Europe, precisely because even a truncated Soviet Union, with its nuclear weaponry, will still retain the largest land army on the Continent. On the other hand, as European economic and political unity grow, there will also be new calls for closer European defense cooperation. So once again the questions are: What is the mission of a European security organization? Who should belong to it? How do the Europeans keep the United States committed to the defense of the Continent? How can German nationalism be contained? And how are the newly pacific Russians to be satisfied that such an organization is not directed against them?

Collective security, the great ideal at the close of World War II, fell apart under the strains of the Cold War. Churchill's "grand alliance," meant to be the cornerstone of collective security, became two alliances, one on either side of the iron curtain. Moreover, these alliances were dominated by two military superpowers: it was not so much collective security as it was American and Soviet military protection that characterized the postwar world. Now it is time to give collective security in Europe a chance once again. It is not enough to have a unified Germany become part of NATO (except perhaps as an interim measure); rather, what is needed is a broad collective security pact among all twenty-three members of NATO and the Warsaw Pact.

Nor should such an organization necessarily come under the aegis of the Conference on Security and Cooperation in Europe (CSCE), the so-called Helsinki process. While the CSCE has been useful in monitoring human rights, its thirty-five-nation membership is simply too large to make it an effective military organization. And while the CSCE, which includes all of the European nations plus the United States and Canada, might be usefully strengthened through the creation of a permanent secretariat to help secure the rights of national minorities and to provide a forum for settling political disputes, its current rule requiring consensus on all actions taken prevents it from imposing sanctions on any of its members; and it is hard to imagine easily changing this rule, for example toward majority voting, when it took the European Community twenty-five years to agree on limited majority voting, and even that still excluded security issues.[3]

A more promising mechanism that might serve as the genesis of a European security organization would be the Western European Union, an organization that came into being as a result of the agreements permitting the Federal Republic of Germany to join NATO after the collapse of the European Defense Community. The Western European Union (WEU) eventually came to include the Benelux countries, Britain, France, Italy, and the Federal Republic of Germany and was formed as a general collective security alliance "to promote unity and to encourage the progressive integration of Europe." Princeton's Richard Ullman has put forth a remarkably persuasive argument for using the WEU to become the charter of a European security organization.[4] The membership of the WEU would be extended to include NATO, the Warsaw Pact, and neutral and nonaligned states that might want to participate, as well as the United States and the Soviet Union. Unlike the

CSCE, the WEU provides for limited supranationality, and the WEU's decisions require simple or weighted majorities, not unanimity. Both Kohl and Mitterrand have suggested that the WEU could become the pillar of a joint European security and defense policy.

The evolving European security organization, however, need not derive from an already existing model. It could as well spring full-blown from the exigencies of the twenty-first century, as NATO itself did in 1949. History teaches us that the notion of trying to preserve military alliances that have outlived their usefulness by giving them a spurious political role never works. A NATO committed not to deploy its forces in eastern Germany, for example, will be of little if any military value once Soviet forces have left Germany. The Warsaw Pact has formally ceased to exist. Nonetheless, countries of the former Soviet bloc will doubtless try to seek some arrangements to provide for their own security. Under these circumstances, the last thing the United States should want to see are ad hoc arrangements among the European countries, which could lead to a neutralized Germany and a French-British nuclear entente.[5]

Above all, a pan-European security system will keep the United States tied to the European Community, even though there will doubtless be economic strains between North America and Europe, with Washington stripped of its customary power to exert leverage over Europe as it once did because of its dominant security position. With the likelihood that U.S. troop strength in Europe will drop from roughly 300,000 in 1990 to something in the order of 50,000 before the end of the decade, the danger for the Europeans, as well as for Americans, is that the U.S. Congress will decide to withdraw all its troops and let the Europeans fend for themselves. This becomes even more likely if America and Europe engage in trade wars and the United States continues to run trade deficits close to $100 billion and fiscal deficits at more than twice that figure.[6]

Moreover, a European security organization that includes the United States and the Soviet Union will help to ensure a European balance of power. It will act as a counterweight to German power and to a regressive Soviet Union; it will effectively tie the United States indefinitely to European security by not asking Washington to bear too great a financial burden. The costs of maintaining some 50,000 U.S. troops on the Continent are negligible, and the U.S. nuclear deterrent will act as a further guarantee of our commitment to European peace and security.

Of a proposed American defense budget for 1990 of about $300 billion, some $150 billion was devoted to sustaining the NATO commitment.[7] Defense expenditures by France and West Germany, on the other hand, total only a little over $60 billion a year. In 1988, the United States spent 6.1 percent of its gross domestic product (GDP) on defense while NATO-Europe spent an average of 3.3 percent. Yet American per capita GDP in 1987 was about equal to that of France, and inferior to the level of Norway, Denmark, Luxembourg, and West Germany, all of which devoted far less of their GDP to defense than did the United States. (Another way of looking at this is to examine total U.S. defense expenditures in 1988, which equaled $820 per capita, whereas in Western Europe the average was approximately $450 per capita, almost half that of America's.)[8]

Reducing U.S. troop strength to 50,000 troops plus support personnel (or two divisions) will save a considerable amount of money. Moreover, a truer figure for totaling up American personnel in Europe is to count the dependents that accompany the 300,000 American troops now on duty; this brings the figure up to 635,000 U.S. military personnel and dependents assigned to fourteen NATO countries. The twenty-fifth largest school district in the United States is located in West Germany. This, too, is changing with the end of the Cold War.[9]

In the past, European leaders were usually outraged at any suggestion that the Americans should draw down their forces in Europe. A unilateral U.S. troop withdrawal would have been labeled destabilizing; it would also have meant that the Europeans would have had to pay more for their defense if they wanted to maintain the same level of mobilization. But with an agreement on a balance of conventional forces from the Atlantic to the Urals, George Bush was able to propose in the spring of 1989 that both superpowers should cut troop strength in Europe to 270,000 each, a formula that would demobilize some 30,000 Americans and 325,000 Soviets. Of course, this would have alleviated any European fears that the West Europeans would have to increase expenditures for their own defense, for they would no longer be left vulnerable to Soviet conventional superiority. Moreover, in the conventional arms talks, ceilings on combat aircraft, helicopters, tanks, and artillery would also have to be achieved. In short, a meaningful agreement had to be across the board, not simply a superpower equation on numbers and deployments of troops.

All this, however, came before the collapse of the Soviet bloc in Eastern Europe. After the 1989 revolution, troop levels such as Bush and Gorbachev had previously agreed on were almost certainly too high. Now the issue was to devise a security system that would allow *any* American and Russian troops to remain west of the Soviet borders.

In this respect, a European security organization would permit U.S. and Soviet troops to remain in Europe, not as occupiers but as guarantors of the European order. It would have as its central mission the assurance that aggression by any country against its neighbor would be met with a united reponse from the organization as a whole. This would radically change the nature of European politics, which, for the past forty years, accepted Russia's right to intervene in Eastern Europe. On the other hand, while a broad alliance does not preclude disputes between members, it could prevent conflicts, such as we have seen between Greece and Turkey (both members of NATO), and potentially between Hungary and Rumania.[10] In addition, tying Soviet military forces to a broader European organization would inhibit the Soviet military from acting on its own across the new borders of the U.S.S.R.

For Germany, as for the United States and the Soviet Union, membership in a European security organization would ease fears that a united Germany would become a source of instability in a fragmenting Europe. In this respect, as the division of Europe ends, the German question will finally be answered.

In any future European security organization, the Europeans, not the Americans and certainly not the Russians, should take the leading role. Yet NATO's supreme commander is always an American, as is the Warsaw Pact commanded by a Russian. This would have to change, and the most desirable candidate for supreme commander, at least at the outset, would probably be a French general as a signal of France's new willingness to join a pan-European military organization. For the United States and the Soviet Union, membership in an ESO would give them a special role in the search for global security, since both are Asian as well as European powers. It would make it less costly for the United States to dispatch troops to other parts of the world, such as the Persian Gulf, if its vital interests—and those of the West as a whole—are threatened. Moreover, admitting the U.S.S.R. into a pan-European defense system would make Soviet participation in the European Community more plausible over time.[11]

The primary purpose of an ESO would be to preserve the territorial status quo, except where changes are mutually agreed on by both parties. Borders are not carved in stone, but the greatest danger that faces the new Europe would be a rectification of borders by force rather than by negotiation. The Rumanian-Soviet border, the Polish-German and Polish-Soviet borders, and the Hungarian-Rumanian border all could easily come under dispute. Under a European system of collective security, however, the chances are good that these could be settled by diplomacy rather than force. In addition, an ESO would monitor arms-control agreements and set up verification procedures, expand conference diplomacy, and create buffer and neutral zones by setting up peacekeeping machinery.

Of central importance is the question of what role nuclear weapons will play in a European security organization. The very fact of U.S. and Soviet participation means that nuclear weapons will continue to guarantee the security of the Continent. No potential aggressor, from any part of the globe, could rule out the possibility that nuclear weapons might be used in retaliation. Moreover, both France and the United Kingdom possess nuclear weapons, and the quantity and quality of these armaments might well increase.

Until now, the cornerstone of NATO strategy has been "extended deterrence," or the willingness of the United States to use nuclear weapons should the Soviet Union have attacked Western Europe with enough force to render a conventional Western defense inadequate. Certainly, the Europeans have insisted on constant assurance that the American nuclear guarantee to defend Europe would hold. In other words, would America risk the destruction of Chicago for Hamburg? This was, of course, an unanswerable question short of nuclear war. Yet the very uncertainty of the response made the American deterrent credible. What Soviet general would have risked the destruction of Leningrad in order to test the validity of the U.S.

guarantee? As the British strategist Sir Michael Howard noted: "If there is one chance in a hundred of nuclear weapons being used, the odds would be sufficient to deter an aggressor even if they were not enough to reassure an ally."[12]

But deterrence does not require sustaining high American troop levels, nor does it call for ever greater numbers of U.S. warheads to be positioned on European soil. "The enduring effectiveness of the American guarantee," wrote former national security adviser McGeorge Bundy, "has not depended on strategic superiority. It has depended instead on two great facts: the visible deployment of major American military forces in Europe, and the very evident risk that any large-scale engagement between Soviet and American forces would rapidly and uncontrollably become general, nuclear, and disastrous."[13] Now, however, an era of conventional force equality is in the offing. The Soviet Union is a potential ally rather than a potential enemy. Under these circumstances, battlefield nuclear weapons are no longer needed—as Bush and Gorbachev have recognized.

In a Europe denuded of battlefield nuclear weapons and with the two superpowers, plus a nuclear-armed Britain and France, linked into a single security system, we are finally left with a policy of *minimum deterrence*. What both superpowers will continue to insist upon are enough strategic nuclear weapons to allow them to retain a second-strike capability. This in itself would serve as an effective deterrent to nuclear war. It would ensure that anyone who dared a nuclear strike would suffer horrifying destruction, but it would not threaten the survival of civilization itself.[14]

With the conclusion of phase I of the Strategic Arms Limitation Talks (SALT) between Washington and Moscow, however, one could envisage a series of negotiations that would go far beyond the first stage, in which the superpowers have agreed to reduce their warheads by one-third. This would still leave the United States with 6,000 and the Soviet Union with 7,000 long-range nuclear warheads. A second stage would be to cut that number by half. And a third and final stage could see the reduction of nuclear strategic warheads to something on the order of 1,000 or fewer on each side. (To arrive at this figure, reductions could be made both unilaterally and through negotiation.)

Once national policy rests on the conviction that nuclear

weapons are not weapons that can be used to deter conventional wars, then this approach seems feasible. It may be impossible to eliminate nuclear weapons absolutely, because the technology exists that makes it possible for other nations to create a nuclear stockpile. But should the superpowers be able to reduce their stock to less than 1,000 warheads each, then truly one could say that the arms race has been reversed and that nuclear strategy as such has become obsolete.

When discussing scenarios for deep cuts in nuclear weapons, it is also important to note that France, Britain, and China will finally have to be brought into the negotiations. Yet even if the three lesser nuclear powers were allowed to retain strategic nuclear weapons at roughly the same level as those of the greater military powers, the United States and the Soviet Union would still enjoy a unique status because of the troops they would doubtless continue to maintain in force and deploy abroad. Only if France, Britain, and China began to assume full responsibility for their own defense would the United States and the Soviet Union truly lose their special status.

Accompanying radical reductions in the stock of strategic nuclear weapons should be a comprehensive test ban, restrictions on flight testing of missiles, and strategies for mutual reassurance to reduce fears of preemptive strikes. The last nuclear weapon will only disappear after it has come to be seen as useless for any purpose whatsoever, and probably long after its very existence has been forgotten.

So much of nuclear deterrence depends on uncertainty. Neither side can really know what the other side will do if nuclear wepaons are ever used. "Existential deterrence," in McGeorge Bundy's phrase, prevails. What this means is that "deterrence is inherent in what any nuclear power can do if it chooses, and also in the possibility that open warfare may drive one side or both to permit in anger what it would never consider in peace."[15] The 1,000 or so strategic nuclear warheads I have suggested both superpowers retain would be ample to support this new form of minimum deterrence.

Under the most optimistic scenario, we can imagine a Europe with a conventional balance of forces, shorn of battlefield nuclear weapons, and backed up by the American *and* Soviet nuclear guarantees, though with both powers retaining a low level of nuclear warheads.

This is a Europe in which the Soviet threat as we have known it has disappeared, and external threats to the peace of Europe are likely to come in the form of terrorism or peripheral skirmishes. Nor are these dangers likely to lead to large-scale war, because the nations of Europe, acting in concert as they did in the forty years after the battle of Waterloo, appear determined to avoid military conflicts. The Concert of Europe at that time did not desire to alter the status quo by force, and all sought an international environment that would let the great powers concentrate on their domestic needs. This is as true today as it was then.

A concert-based European security system offers perhaps the most promising future for Europe. Political scientists Charles and Clifford Kupchan remind us that "a concert system revolves around a small grouping of major powers that share similar views of a desirable international order and are therefore able to forge cooperative mechanisms for preserving peace."[16] In such an organization it would be desirable to have at least an informal security group that would consist of Europe's major military powers—Britain, France, and Germany—as well as the United States and the Soviet Union. While this core security group would probably retain no explicit decision-making rules, the group would be expected to allow the common interest in preserving the peace of Europe to prevail over short-term interests.

Whether or not a core security group comes into being, a Europe that acted in concert to restrain aggression is surely the Europe the United States hoped for when it articulated its doctrine of containment almost half a century ago. To preserve the balance of power in Europe under new conditions and with a new policy of including the Soviet Union rather than excluding it from collective security arrangements should now be the American goal.

But even as America helps to devise organizations that will lead to a peaceful Europe, its role will necessarily be reduced. Not only will the U.S. troop presence shrink, but it will be the Europeans, more prosperous than ever, who will take the lead in developing the institutions of the European Community that will ultimately provide Europe with greater military and political weight than it has known for the past half-century. In many respects the United States will play a somewhat marginal role in the politics of Europe, precisely because there are institutions already in place that will give Europe greater security and a stronger voice in foreign affairs.

In East Asia and the Western Pacific, on the other hand, the United States assumes a very different function. There are no military and political organizations, no economic Common Market, to bind those nations together. There is no one to replace the United States as the guarantor of security in the Western Pacific. Yet the economy of the region is the most dynamic in the world, dominated by the financial power of Japan. It is also shadowed by the colossus of a restive China on the verge of political turmoil, and a weakened Soviet Union attempting to pursue a far more flexible diplomacy than it ever has in the past. In the Pacific Century, if that is what it proves to be, the United States should be determined to play a central role in a region that may well have a greater impact on America than U.S.–Soviet relations or the new European order.

PART III

Pax Pacifica

7

Asian Insecurities

For forty years American military strategy has focused on the European central front and the need to defend the West against Soviet forces in East Germany and Poland. Yet the only large-scale wars the United States fought during this period were in Asia. American military planners remain haunted by the memory of Pearl Harbor and the fear that an enemy, even in the age of nuclear deterrence, might someday launch a surprise attack. The need for the United States to rethink its security requirements in the Western Pacific, however, has grown even as American economic preponderance has lessened. No longer can the United States base its military planning on the containment of the Soviet Union.

East Asia's growing economic strength came about in no small part because the United States was so focused on its security concerns that it subordinated its economic interests to its geopolitical needs. Under the exigencies of the Cold War, Washington actively promoted the economic recovery and growth of Japan, Taiwan, and South Korea. America was willing to keep the U.S. market relatively open while those of its East Asian allies would remain much more restrictive. In so doing, America provided for the security of its allies while encouraging them to become ever stronger economically. This may have been a wise policy during the heyday of the Cold War. But Mikhail Gorbachev's rise to power has made it possible for Washington to reverse its priorities and put its economic needs to the forefront. Not only can the East Asians now bear the brunt of their own defense, but both they and the United States can now seek substantial reductions in overall defense spending as the Soviet threat ends.[1] Under these changed conditions, both East Asia and America should be able to prosper. Yet while the Soviet Union has set forth a new diplomatic strategy aimed at reducing military and naval deployments in the region and is seeking an enlarged economic and political role in East Asia, Washington has seemed bound to a policy of preserving the status quo. The new challenge for Washington is to replace its geopolitical emphasis on anti-Sovietism with a new focus on geoeconomics.

American presidents fought both the Korean War and the Vietnam War to contain what they believed were the expansionist aims of the Soviet Union and China. But by ignoring the growing rift between Moscow and Beijing in the mid-1960s, the Kennedy and Johnson administrations may have actually contributed to Sino–Soviet solidarity. On the other hand, by recognizing the degree to which Beijing and Moscow could be separated, the Nixon administration exploited the rift between the two big communist powers, which may have helped bring about the 1973 Paris accords terminating the formal American involvement in Indochina. By the 1980s, all the big powers in the region believed they needed the United States to maintain a balance of power in Asia. The "stable equilibrium" among the United States, Japan, China, and the Soviet Union, which Secretary of State Cyrus Vance urged when Washington established full diplomatic ties with Beijing in 1979, seemed in the offing at last.

With the death of Mao Zedong in 1976 and Leonid Brezhnev less than ten years later, the Sino-Soviet rivalry, heightened by military buildups on both sides, sharply diminished. The summit meeting in May 1989 between Mikhail Gorbachev and the aging reformist Deng Xiaoping signaled not so much a rapprochement as a recognition that neither country should pose a military threat to the other. At the time, both nations were led by men trying to reform the centralized economic systems they had inherited, desperate to become competitive in a world that was fast condemning them to permanent backwardness. Both nations had also suffered military setbacks in their aims to impose their wills over their neighbors—China with its 1979 invasion of Vietnam, and Russia with its decade-long intervention in Afghanistan. Both nations, however, soon found themselves consumed by internal tensions—the Soviet Union because of the independence movements within its republics, and the Chinese in the wake of their murderous crackdown on students demonstrating for reform and democracy in June 1989.

By then, Japan, a once mighty military rival that had humiliated both Russia and China earlier in the century, was now spending more for defense than any other country in the world except for the United States and the Soviet Union. Its gross national product was second only to America's. And no country was spending more on foreign aid. It is the region's biggest investor and banker. But while Japanese investment is welcomed in China, South Korea, and the littoral states of the region, Japanese economic and military domination is still feared by those same nations. The last thing Beijing wants—or, for that matter, the smaller states on the periphery of Asia stretching from Korea to Thailand—is a latter-day version of Tokyo's wartime "Greater East Asia Co-Prosperity Sphere." Fearful of too great a dependency on the Japanese economy, Deng Xiaoping said in 1979 that "America's trade with China must come equal to Japan's."

All the great powers desire the United States to play a major role in the Far East. As Gorbachev said, "The United States is a great Pacific Ocean power. . . . There is no doubt that without the United States, without its participation, one cannot solve the problem of security and cooperation in the Pacific Ocean zone in a manner satisfactory to all the states in the region."[2]

The idea of a four-country balance of power in the Far East was already present during the days of the Nixon–Kissinger partnership. "We must remember," Richard Nixon said in 1972, "that the only time in the history of the world that we have had any extended period of peace is when there has been a balance of power. . . . I think it will be a safer world and a better world if we have a strong, healthy United States, Europe, Soviet Union, China and Japan, each balancing the other, an even balance."[3] At the time Nixon spoke, the world he evoked was more prophecy than reality. In East Asia and the Pacific, however, a four-power balance was already coming into being. Nonetheless, stability in the region is far from assured. The breakup of the Soviet Union, the tensions within China that could come to the surface when an aging leadership gives way to a new generation, the economic hegemony of Japan—any of these developments is likely to lead to instability in the future. It is precisely in order to guard against such instability that the United States must continue to play the role of the balancer.

Economic interests alone dictate a central role for the Far East in American policy. Trade flows to the Pacific region have shifted dramatically over the past quarter of a century. U.S. trade with East Asia has tripled over the past decade. For example, in 1987, America's trade with the region amounted to 37 percent of its total trade, whereas its trade with the European Economic Community came to only 20 percent. And the upward trend continues. From 1980 to 1984, the Pacific Basin's share of U.S. exports rose from one-fourth to one-third of the total. Japan counts as the second largest U.S. trading partner after Canada. Its imports from the United States in 1989 came to $44.5 billion; 60 percent of those U.S. sales were manufactured goods, higher than U.S.-manufactured exports to Germany and France combined. Yet well over half of the U.S. trade deficit rested with the Pacific Basin.[4] Indeed, by 1989, out of a $109 billion U.S. trade deficit, Japan ($49 billion), Taiwan ($13 billion), and South Korea ($6.3 billion) accounted for $68.3 billion, or nearly two-thirds of the total.[5] As East Asia's share of the world product more than doubled in the twenty years from 1967 to 1987,[6] it is hardly astonishing that one economic expert predicts that this area will account for 50 percent of the world product by the turn of the century.[7]

The financial prowess of East Asia also continues to grow. Of the world's ten biggest commercial banks, seven are Japanese. In 1986–87, Japan's banks provided 16 percent of the foreign capital the United States needed, and together the United States and Japan accounted for 70 percent of total foreign investment in the states of ASEAN, the Association of Southeast Asian Nations (Thailand, Indonesia, Malaysia, the

Philippines, Singapore, and Brunei).[8] Indeed, between them the United States and Japan account for some 55 percent of the world's banking assets.[9]

The security interests of the United States in East Asia and the Western Pacific rival America's economic entanglements. Despite a foreign debt approaching $700 billion in the 1990s, the United States still stations some 135,000 American troops in the Western Pacific, at a cost of $30 billion a year.[10] Nearly half of the strength of the U.S. Navy—with 33,000 sailors afloat—two-thirds of the combat wings of the U.S. Marine Corps, two army divisions, and several fighter wings of the U.S. Air Force make up the American commitment there. The Pentagon plans to reduce this number to 120,000 by 1993.[11] (This costs far less than U.S. military expenditures for NATO, which run to about $150 billion a year and whose reductions will not make a serious dent in the defense budget until the mid-1990s.)[12]

Despite Soviet initiatives on lowering military confrontation in East Asia and the Pacific, the Bush administration continued to call for a grouping of states, ranging from New Zealand to China, united "in order to contain Soviet expansionism in the area."[13] As it is, the United States has ringed the Western Pacific through bilateral treaties with Japan, Korea, and the Philippines. The Manila Pact adds Thailand as a treaty partner, and finally there is the treaty with Australia and New Zealand. Jerry Sanders, the director of Peace and Conflict Studies at the University of California at Berkeley, has pointed out that Bush justified this quasi-alliance system by claiming that the Soviet Pacific fleet was "already larger than the entire U.S. Navy."

This just isn't so. Not only are the Soviets hemmed in by superior U.S. naval deployments along the perimeter from the Philippines to Japan, but "the Soviets have actually reduced the total size of their naval deployments in the region and have called for negotiations to lower the levels still further."[14] From Indochina, about 80 percent of Soviet air and naval units—and all combat aircraft—have already been withdrawn.[15] In both economic and military terms, then, the American commitment to the Far East is substantial, yet there is little evidence that the Bush administration has begun to react to the changing nature of this commitment in the light of Soviet foreign policy initiatives or the growing economic power of Japan.

American foreign policy has continued on the same path it had mapped out some forty years before—providing for Japanese security and urging the Japanese to share increasingly more of the military burden. And so, while Washington spent over 6 percent of its gross national product and 70 precent of its federally funded research and development for military purposes, Japan spent a little over 1 percent of *its* GNP for defense and almost all of its research and development funds for commercial innovation.[16] Yet it seems almost impossible to imagine a situation in which Japan will go on being willing to act at the behest of her protector by fulfilling the defense mission assigned to her—which now includes responsibiliity for air- and sea-lane defense up to 1,000 miles from her coastline and up to 500 miles on either side of the Soviet port city of Vladivostok.

The Japanese defense budget (in dollar terms) is already the third largest in the world.[17] If Japan is, as former Prime Minister Tashiro Nakasone called it, an "unsinkable aircraft carrier," it may well consider that playing a central defense role means that it should also have a greater say in who defines that role.[18] It would indeed be a bizarre turn of history if Japan stays content to remain a junior partner of the United States even as it assumes the role of America's largest creditor.

Nonetheless, the American military presence has provided the United States with a degree of leverage it might not otherwise have. Any country that relies on the U.S. military commitment cannot blatantly ignore America's complaints about its economic growth. In the 1980s, for example, Japan had decided to develop its own state-of-the-art fighter jet, to be known as the FSX. The United States very vocally and publicly protested, because it feared that the Japanese would overtake America's lead in aerospace technology. Finally, despite the importance the Japanese had attached to an independently developed fighter, they agreed to joint development with McDonell-Douglas based on its F-16.[19]

For the United States in the past four decades of the twentieth century, the strategy of the containment of the Soviet Union and a commitment to free trade have been the cornerstones of American policy toward Asia. Unfortunately, American economic profligacy has put the United States in the awkward and ultimately untenable position of having to borrow from those whose very security it has been committed to protect. Moreover, under Mikhail Gorbachev, the Soviets have proposed striking new initiatives in the Far East that may produce a less menacing Soviet military presence and therefore more restive U.S. allies who find the postwar U.S. security system no longer relevant to their needs and the doctrine of containment the victim of its own success.

8

Across
the Pacific

VLADIVOSTOK has long been the symbol of Russia's Pacific presence. Soviet battle groups operate out of the port city, and Vladivostok remembers that Japan annihilated the Russian fleet in 1905 during the Russo-Japanese War. It was therefore appropriate that Mikhail Gorbachev should choose Vladivostok to deliver a speech signaling a new Russian policy toward Asia. On July 28, 1986, he called for reductions of all navies in the Pacific Ocean, the general reduction of armed forces and conventional arms in Asia, and "confidence-building measures" for the security of Pacific sea lanes, as well as steps to prevent international terrorism. These proposals were linked to his suggestion for a regional conclave, modeled on the 1975 Helsinki Conference on Security and Cooperation in Europe, "of all countries having a relationship with the ocean."[1] In addition, Gorbachev offered to thin out troop concentrations on the Sino-Soviet border, and he suggested that talks should take place with Japan over the Soviet seizure of Japan's "northern territories" at the end of World War II.[2]

Gorbachev further proposed eliminating medium-range missiles deployed in the Far East without linking this initiative to the U.S. nuclear presence in South Korea, the Philippines, and the island of Diego Garcia. After the conclusion of the intermediate-range missile talks with the United States, in the treaty signed in December 1987 in Washington, all medium-range missiles were indeed scheduled to be dismantled.

Two years later, in a speech delivered in the Siberian city of Krasnoyarsk, the Soviet leader proposed a series of confidence-building measures to further enhance security in Asia and the Western Pacific. He also called for six-nation talks on lowering the level of air and naval activity in Northeast Asia, and for what he described as a negotiating mechanism to consider Asian security proposals among the United States, the Soviet Union and China as permanent members of the U.N. Security Council.[3]

Then, in an address at the United Nations in December 1988, Gorbachev announced a Soviet unilateral withdrawal of 500,000 troops and the scrapping of 10,000 tanks (5,300 of which would be of the most advanced design).[4] Of particular significance for Gorbachev's Asian policy were reports that 200,000 troops would be removed from the Asian theater, along with three-fourths of the approximately 55,000 ground troops in Mongolia, as well as all Soviet air force troops there. In short, proportionately, the biggest of the military cuts were to come not from Europe but from Asia.

At sea, Moscow cut the size of the Soviet Pacific naval fleet by forty ships between 1984 and 1988, and began to withdraw its naval forces from Cam Ranh Bay.[5] In fact, the Soviets have not had a "huge" naval base at Cam Ranh Bay.[6] For if Cam Ranh, which once had 2,500 Soviet personnel and three to four major surface combatants, was huge, Guam, with 9,100 U.S. personnel, and the Philippines, with 16,400 U.S. soldiers and sailors, was humongous. In any case, the Soviet military presence in Indochina was fast disappearing.

Despite widely held assumptions in the United States that the Soviets have been engaged in a major military buildup in the Far East, the figures tell a different and more encouraging story. According to the International Institute for Strategic Studies in London, in 1980 the Soviets deployed eighty submarines, eighty-six major surface combatants, and fifty-four amphibious craft. In 1987, they deployed only seventy-six submarines, eighty-two major surface combatants, and twenty-one amphibious craft. This shows an overall *decrease* in force levels.

The only significant increase was in ballistic missile nuclear sub-marines—from twenty-four in 1980 to thirty-two in 1987. While this could affect the region's strategic balance, it does nothing to increase the Soviet Union's capability to wage war.[7]

Washington was hardly responsive to Gorbachev's initia-tives. In March 1990, Defense Secretary Dick Cheney offered to reduce American forces in Asia, about 120,000 troops, by approximately 10 percent.[8] This excessively modest reduction was done purely for budgetary reasons and showed no disposi-tion by the Bush administration to rethink its Asian security needs in the light of a receding Soviet threat. Washington has been especially unwilling to negotiate naval deployments with the Soviets, even though high-ranking Soviet officials have pointed out that deep cuts in conventional and strategic nuclear forces in Europe will someday have to be linked to reductions in the naval deployments of both powers.

An unintended consequence of the Soviet buildup in ballis-tic missile nuclear submarines, as Admiral William J. Crowe and Asian expert Alan D. Romberg suggest, has been to stimulate an American response in kind. To reduce the threat that each side perceives, they propose "agreement on substantial cuts in attack submarines." In addition, "there would be substantial benefits from mutually denuclearizing all naval forces except for sea-launched ballistic missiles," which fall under the rubric of the START agreement on long-range strategic missiles.[9]

At the very least, the United States could begin to examine confidence-building measures with the Russians, as we have done in Europe: notification of major naval exercises, invita-tions to observers, dialogue between senior officers on military doctrine and deployments. Washington could also move to reduce the number of sea-based missiles as part of an overall reduction in strategic weapons, something Bush pointedly refused to do when he offered unilateral reductions in nuclear weapons in September 1991.[10] The essential issue is that Wash-ington has made discussion of any serious cuts in its naval forces off-limits. Yet there is no reason to do this. A wise policy would be to seek a lowering of military tensions in the Pacific through bilateral talks with the Soviets and, in so doing, encourage the Japanese and the Chinese not to embark on greater military buildups of their own. None of this implies a U.S. withdrawal from the Pacific.

Although the Soviet Union offered proposals for arms reductions throughout Asia and the Pacific, Moscow was especially keen to repair relations with Beijing. Ever since the Soviets and the Chinese broke off relations in 1963, both countries had built up their armed forces along what has been the longest disputed border in the world. From a low of between fifteen and twenty divisions deployed along the border in 1963, the Soviets steadily increased their forces to a high of fifty-six divisions in the 1980s. All this has now changed. Moscow withdrew about 100,000 men before Gorbachev's historic meeting with Deng Xiaoping in May 1989, and Soviet troop deployments are likely to be reduced further—to at least their pre-1963 level of fifteen divisions.[11]

For the Soviet Union, closer relations with China have depended on meeting three conditions insisted upon by Beijing: that the border dispute between Russia and China be settled; that Russian forces withdraw from Afghanistan; and that the army of Russia's client-state Vietnam withdraw from Cambodia. By the time Mikhail Gorbachev disembarked at Friendship Airport in Beijing on May 15, 1989, all three conditions had been essentially fulfilled, and the Chinese leadership's subsequent brutal crackdown on student and worker dissidents has not affected these understandings.

The Russians agreed that the frontier should pass along the "main channel" of the Amur and Ussuri rivers, which meant dropping the claim that the border should run along the Chinese bank. This concession might allow China to appropriate as many as 600 small islands. In Afghanistan, the Soviets withdrew all of their troops by mid-February 1989, which seemed to mark the end of Russia's military adventurism outside its own borders. Under Soviet prodding, the Vietnamese withdrawal from Cambodia continued throughout 1990, but the future of Cambodia seemed less assured, with factions of the murderous Khmer Rouge, supported by Beijing and whose guerrilla army numbered some 35,000, contending with a Vietnamese-backed government, while Prince Norodom Sihanouk struggled to maintain a government controlled by neither. Yet even in Cambodia, prospects for peace grew in 1991 when the U.N. Security Council endorsed a plan to have the United Nations monitor a ceasefire, supervise a transitional government, and help prepare and monitor free elections;[12] later that year, Cambodia's warring factions endorsed the plan, reduced their armies by 70 percent, and put the rest under U.N. supervision.

Despite the internal problems that so consumed the Chinese leaders, their efforts to repair and deepen relations with the Soviet Union did not significantly diminish. But they were wary of the turmoil they saw spreading within the Soviet Union. As China's prime minister, Li Peng, said in Moscow in April 1990, the Soviet model of political and economic change did not apply to China.[13]

While trade between China and the Soviet Union is small, it once seemed to promise greater rewards for both countries. Sino-Soviet trade increased from $300 million in 1982 to $2.6 billion in 1986. Moreover, China, which was industrialized with the support and advice of the Soviets in the 1950s, can use Soviet industrial expertise to modernize its own factories. The Russians, in turn, can make good use of the raw materials and inexpensive light manufacturing goods available from China. It is also important to remember that Beijing does not want to become too economically entangled with Japan. This was an added incentive for China to seek closer economic ties with the Soviet Union, as it did prior to the Beijing massacre.

By 1990, the deepening Soviet economic problems that followed Gorbachev's efforts to restructure the economy gave the Chinese, with their own stagnating economy, little hope that they could find much relief in Soviet trade and investment. While neither Beijing nor Moscow expects a return to the hostility of the 1960s and 1970s, any deeper rapprochement between the two powers is also unlikely, at least in the near term. The Chinese leadership fears any economic reform program that might lead to a transition to democracy. Above all else, they fear chaos. The spectacle of a disintegrating Soviet Union and especially the dismantling of the Communist party are simply anathema to Beijing.

If the Soviets moved rapidly to repair relations with Beijing, they were stymied in their efforts to draw closer to Tokyo. The Russian occupation after World War II of Japan's Northern Territories, the four southernmost islands of the Kurile chain, has never been accepted by Tokyo. As a result, there has been no final peace treaty between Moscow and Tokyo, although diplomatic relations were restored in 1956. In the 1970s, when the Soviet Union began to build up its military capabilities in the Far East (which included militarizing two of the Kurile Islands, Etorofu and Kunashiri, where they deployed a full division of troops and twenty MiG-23 fighters), Japan responded by reevaluating its own security policy. The presence of two small Soviet aircraft carriers, several modern cruisers, and amphibious assault ships in the area, plus the deployment of SS-20 missiles east of the Urals in the early 1980s, helped persuade Japan to establish even closer strategic ties to the United States. If anything, these Soviet moves stimulated the Japanese to increase their defense budget, pushing it over the limit of 1 percent of GNP that had been traditionally observed.

Gorbachev, however, indicated a new flexibility in the Soviet approach to the Northern Territories. On a visit to Tokyo in September 1990, the Soviet foreign minister said Moscow was willing to negotiate the four-decades-old territorial dispute. Eager for Japanese trade and technology, Moscow may well decide to substantially reduce Soviet ground and air forces in the Kuriles and redeploy an aircraft carrier and an amphibious troop carrier. Some demilitarization of the Northern Territories, perhaps accompanied by an offer to turn them into special economic zones, or even the return of two of the islands, would be significant moves toward normalization of relations between the two great powers and a final peace treaty. Moreover, given the loss of Eastern Europe and independence for the Baltic states, the Soviets may even feel that giving back the Kurile Islands no longer provides a dangerous precedent for yielding up territory they controlled after World War II.

Yet there was no significant movement on the issue of the Northern Territories on Gorbachev's visit to Tokyo in May 1991. He may well have been hobbled by his own military in his desire to show some genuine Soviet give on the territorial dispute. In any case, the visit accomplished very little, and he returned to Moscow without any likelihood of significant Japanese investment in the Soviet Union. Nonetheless, after the August 1991 Russian revolution, a high Soviet official endorsed Japan's claim to the islands.[14]

In the 1990s, the Soviet approaches to the United States, China, and Japan must be evaluated in two ways. Moscow clearly wants to get into the great economic game in Asia. Already the Soviets have sent observers to the Asian Development Bank and are seeking to become members. But the Soviet

Union's current stake in the Asian economy is very limited. Asia accounts for only 3.3 percent of all Soviet trade.[15] To change this Moscow must heal its sick economy and repair its political ties to the other powers in the region.

In this respect, Soviet relations with South Korea have dramatically improved. Following Gorbachev's ill-fated visit to Japan, the Soviet leader proceeded to South Korea where he and President Roh Tae Woo agreed to multiply their trade tenfold over the next four years. They also called on North Korea—not so long ago Moscow's key ally in the region—to open its nuclear installations to international inspection; and Gorbachev endorsed South Korea's campaign to gain membership in the United Nations. Gorbachev aptly summed up the difference between his reception in Japan and in South Korea: "It was cold with rain in Japan, but here [in South Korea] it is so warm I feel right at home."[16]

The central aim of Moscow's Asian policy is to strengthen the Soviet Union's own economy. But for the Soviet Union to enter the Asian market requires Moscow to rethink the very essence of Soviet security. In less than a decade and a half, the Soviet Union has built up its naval presence so that now its Pacific fleet has grown from little more than a coastal defense force into the largest of the four Soviet fleets. Should Moscow further increase its naval forces, or make any significant addition to the 30 percent of Soviet intercontinental ballistic missiles and strategic bombers now deployed east of the Urals, the new Soviet leadership will have failed in its effort to establish an economic bastion in the Asian market. For Russia (as for the United States), there can be no separating its need for security from its desire for trade and investment.

America's task in Asia, so long conditioned by the perception of a Sino-Soviet and later simply a Soviet threat, has to change as the Soviet threat evaporates. While continuing to maintain a significant though reduced military presence, America must play a major economic role, and that means strengthening our economy and devoting our diplomatic energies to our ties—economic, political, and military—to our once great enemy and now our ally and potential economic adversary, Japan.

9

Japan: The Ascendancy of the Trading State

"WE are always being told, 'you should take on more responsibility,'" a former Japanese trade negotiator said. "If we try to raise our voice, we face the comment that 'you are trying to dominate.' If we are silent, we face the criticism that 'you are too silent.'"[1] Though Japan has surpassed the Soviet Union to become the second largest industrial economy in the world and is now the largest donor of foreign aid, Japan's world role, and even its role in Asia, has been defined in Washington.

This is not likely to last much longer. Tokyo's economic weight in the U.S.–Japan connection is too strong to preserve the relationship that the postwar architects of American foreign policy designed. America is Japan's leading trade partner; Japan is America's second biggest after Canada. In 1989, Japanese-American trade showed Japan running a $50 billion trade surplus. With $83 billion in cumulative two-way investments (through 1989), Japanese firms became the fourth biggest foreign direct investors in the United States, while U.S companies remained the biggest in Japan. By the year 2000, Japan, a nation of 120 million people, may produce a GNP only 15 percent smaller than America's.[2] Most significant of all,

Japan has been financing a third of America's yearly budget deficit. Yet despite the outward trappings of equal partnership, Japan has been assigned the role of a very junior partner in the United States–Japan mutual security alliance. The postwar willingness of the Japanese to accept that position also accounts in large part for the singular role Japan plays in the Asian balance of power.

The relationship between the United States and Japan is by far the most important one in the Far East. The United States is Japan's only formal ally, and Japan relies on America to provide its only nuclear deterrent. Yet the continuing disputes over narrow trade issues endanger the fundamental ties between the two countries. As the Japanese have maintained a trade surplus with the United States of some $50 billion per year during the late 1980s, American resentment that this trade balance cannot be righted in America's favor has inevitably affected issues relating to U.S.–Japanese security. The U.S. Congress, for example, has repeatedly called on the Japanese to increase their defense spending so that America can spend less. This short-sighted policy ignores the changing strategic environment in East Asia and the Western Pacific. Yet it is equally dangerous and self-defeating if Americans think that the economic aspects of the Japanese–American relationship can be separated from its political and security dimensions.

The cornerstone of Japanese security was the adoption of the Japanese Constitution of 1946, drafted under the direction of General Douglas MacArthur, which renounced war and espoused an "international peace based on justice and order." But the outbreak of the Korean War significantly changed the attitude of American policymakers toward the role they wanted to see Japan play in East Asia and the Western Pacific. In essence, Washington wanted to erect a barrier against presumed Soviet aggression in the Far East, and to this end Japan needed to rearm. The Japanese did rearm but, when they signed the 1960 mutual security treaty, they also acknowledged their dependence on the United States.

In the era of marked U.S. military superiority over the Soviet Union, especially in nuclear weapons, and with the U.S. economy the locomotive of world trade, Japan not only remained dependent on the United States for its security but also took the lead from Washington on the political role it should play in international affairs. It was not until the 1970s that Japan's stance on military issues began to change. This came about for three reasons: the Soviet buildup in Asia under Leonid Brezhnev, the Japanese belief that American military superiority and interest in Asia had declined in the wake of the Vietnam War, and Jimmy Carter's decision (later rescinded) to withdraw U.S. troops from the Korean peninsula. In 1976, Tokyo agreed to a defense plan with the United States that set forth a new objective—that Japan should be able to defend itself against a small, limited attack until the United States arrived in force to beat back the assault.

The Japanese defense mission—which the Americans have insisted upon—is not only to defend the northern island of Hokkaido, but also to seal up the straits through which the Russian navy would have to pass on its way from Vladivostok to the open sea. Japan is further expected to mount an air defense against attack from the Siberian mainland and to patrol the shipping lines from Japan as far south as the Straits of Malacca (between Indonesia and Malaysia). Yet the U.S. Congress has continued to call for Japan to bear an ever greater share of the defense of Asia, though always under U.S. tutelage. While the United States spent 6.5 percent of GNP (1987 figures) on defense, Japan spent only 1.004 percent; in 1990–91, the projected Japanese budget was .095 percent. In dollar figures this means that Japan was spending about $30 billion, when the United States was spending about $300 billion. Because of the appreciation of the yen, however, this has made Japan the third largest defense spender in the world. (Budgets notwithstanding, most experts rate Japan the world's sixth or seventh power in actual military capabilities.)[3]

As it is, while the United States deploys almost twenty-five destroyer-type vessels in the Western Pacific and the Indian Ocean, Japan operates more than twice that number, and that total will climb to at least sixty in the 1990s. Again, the United States maintains about twenty-five P-3 antisubmarine aircraft in the Asian theater; the Japanese have 100 P-2s and P-3s stationed within 500 miles of Vladivostok on a daily basis, and the 300 Japanese fighters exceed those that the United States has stationed in Japan, Korea, and the Philippines combined.[4] In 1990, Japan was to have had 1,205 tanks, 163 F-16 fighter planes, 100 antisubmarine aircraft, and 16 submarines.

Despite Moscow's peaceful overtures, Tokyo expects to continue a relatively high level of defense expenditure through the 1990s: Aegis-capable guided missile destroyers will be purchased, with their high-capacity air defense systems; over-the-horizon radar is to be developed and deployed; and the new F-16A support fighter—intended to be one of the best in the world—is to be acquired. The Japanese have made it clear, in former Prime Minister Norburo Takeshita's words, that they must have "defense capabilities commensurate with [their] national power."[5]

The extent and nature of Japanese national power is precisely what troubles the United States and the other powers of the region. For Washington, it is Japan's continuing trade surplus with America that is the principal source of friction. All U.S. efforts to close the trading gap have failed. From a surplus of $48.7 billion in the year of the 1985 Plaza accord that was designed to drive down the value of the dollar and improve the U.S. trading position, the Japanese surplus just went on rising—to $58.6 billion in 1986, to over $60 billion the following year, and then dropping to $50 billion in 1989.[6]

To counter the Japanese trade advantage, Washington has moved to curb the flow of U.S. imports from Tokyo. Fearful that the Japanese would unload Japanese microchips on the world market and so hinder the export of American chips, the Reagan administration and Congress imposed penalties in 1987 on the Japanese microchip industry and levied 100 percent retaliatory tariffs on selective imports such as color TVs, laptop computers, and power tools. Even the Japanese acknowledged that the huge Japanese trade surpluses were unconscionable and that Japan would have to make some concessions to the United States. Yet once again, Tokyo believed that Japanese hard work, Japanese technological competence, and Japanese managerial skill were being made the scapegoats of America's general unwillingness to compete. In addition, the United States appeared unable to substantially reduce its enormous federal deficit that encouraged Americans to import more than they export.

For the Japanese, the overriding goal is to become technologically superior to the Americans. In the Gulf war, the highest-tech parts of the American systems were often Japanese. Moreover, as Asian specialist James Fallows reports, "the machine tools on which [the U.S.] defense industry relies, the optical elements of precision-guidance systems, the composite materials used to make high-speed aircraft, come increasingly from suppliers in Japan."[7] By seeking technological superiority, the Japanese hope to come up with technological autonomy. If they achieve this, they will no longer be reliant on foreign suppliers for their defense and will be able to provide for their own security.

This is not to say that Tokyo would prefer to do so at this time. The U.S. commitment to sustain a significant air and naval presence in the Western Pacific and the nuclear umbrella that protects Japan have well served Japanese interests. But in the event of a radical U.S. retrenchment in the absence of a palpable Soviet threat, the Japanese do not intend to stand naked before *any* prospective enemy. As a rich country, Japan can be expected to exercise one of the normal attributes of sovereignty—providing for its own security in the absence of a reliable protector. Even today, Japan produces 90 percent of its weapons domestically.[8] In short, if Japan needs to go it alone with conventional arms, it must have the wherewithal to do so. In this respect, Tokyo concluded that it must have an aerospace industry that can challenge America's, not only because aerospace exports would further strengthen the Japanese economy, but also because they would provide Japan with a fair degree of national security.

The FSX affair perfectly illustrates this policy. By the end of the century, Tokyo needs to replace the Japanese-built F-1 fighters. Two U.S. aircraft companies proposed fighters that could easily fulfill Japanese requirements—the F-16C and the F/A-181A. It would have been financially prudent for the Japanese to have bought the F-16, but instead they preferred to construct their own aircraft, the FSX, though that would cost twice as much as buying the F-16s off the rack. Under pressure from the Reagan administration, however, the Japanese finally agreed to give an American company, General Dynamics, 30 to 35 percent of the $1.2 billion cost of the FSX in return for F-16 technology.

Though the costs are high, Japan will gain the experience of putting together a modern aircraft from drawing board to take-off. In short, were the Japanese truly concerned to get the most for their defense yen and reduce America's trade deficit with Japan, they would buy the F-16. On the other hand, if the

Japanese goal is to have a defense capability of its own, then the FSX decision marks a significant milestone on the path to self-sufficiency that may ultimately involve some limited air and missile defense, a Japanese version of former President Reagan's long-sought Strategic Defense Initiative.

At the heart of the Japanese-American disagreements is a propensity on both sides to talk about the bad habits of the other's society. The Japanese feel that Americans simply spend too much. If only the Americans would reduce their budget deficit, offer tax incentives to save more money, spend more on education, and encourage exports, all would be well, the Japanese believe—and they are right. If only the Japanese would make it easier for foreigners to sell their goods in Japan and spend more money on both consumer goods and public works, all would be well, the Americans believe—and they too are right. But the likelihood is that neither country will seriously modify its behavior.

Nor are relations between the two powers improved when Japanese nationalistic assertiveness evokes memories of the Japanese military buildup in the late 1930s. This came to the fore in 1989 when a major Japanese industrialist and a prominent politician wrote a book titled, *The Japan That Can Say "No,"* and approvingly quoted Minoru Genda, the planner of Pearl Harbor—"Japan will be all right. It is able to defend itself. Japan's technology can be the basis for Japan's defense."[9] Similarly, Japanese–American relations are not helped when the U.S. Marine commander in Japan declares that the American military presence there is "a cap in the bottle" to prevent a resurgence of Japanese military power.[10]

What is needed is less contentious bargaining and a more creative U.S. policy that will redefine the nature of our alliance with Japan. In respect to international financial institutions, in view of Japan's role as the world's leading donor of foreign aid, the United States should suggest that a Japanese head the World Bank. As it is, Washington's insistence that its plans for dealing with Third World debt and various development problems are somehow inherently better than anything the Japanese can come up with is both unreasonable and unconstructive.[11]

We should also join together with the Japanese to upgrade the General Agreement on Tariffs and Trade into a full-scale international trade organization that would deal with financial services and technological development. In this way the Japanese would find it politically easier to remove trade barriers and other impediments to international trade and investment.

These examples point up the problem American policymakers have in sharing their decision-making powers with their allies rather than simply asserting their will. Our policy should be to encourage the Japanese to take a more visible place in deciding international issues, and one way to start would be for the United States to actively support a seat for Japan on the U.N. Security Council.

Most important of all, we should stop pressuring the Japanese to increase their defense spending and recognize that the ending of the Cold War means that the Japanese, too, can look for ways to reduce their defense budget. By enlisting the Japanese in talks with the Soviet Union to seek a reduction in military and naval deployments in the Far East and the Western Pacific, the two great allies can work in concert to develop new proposals to lower the level of military confrontation in the region. By testing the sincerity of the Soviet Union, Washington and Tokyo can continue to play their respective roles in maintaining the balance of power in the Far East, but with significantly less emphasis on military prowess. Moreover, in seeking new initiatives to lower tensions in the Far East, Washington and Tokyo must accord China, the third nuclear power in the region, a central place in the diplomacy of post–Cold War Asia, even as they seek to ameliorate China's repression of basic human rights.

10

China: The Power of Weakness

ON May 17, 1989, a million students and workers had gathered in Beijing's Tiananmen Square in perfect discipline to demonstrate for freedom and democracy. Appalled by corruption in high places, the students really were demanding accountability from the political leadership of their country, though they did not use this word. Nor did they demand a multiparty system, as in Hungary and Poland. They even brandished posters of Mao Zedong, who seemed to them the embodiment of incorruptibility rather than the inspirer of the Cultural Revolution. For this was a generation that had never known the horrors of the decade-long Cultural Revolution, which had taken place in the mid-1960s, when the party mobilized the young to overturn the established order.

The demonstrations in Beijing had begun in mid-April to commemorate the death of Hu Yaobang, the former party leader who had been accused by the hierarchy of being too liberal in his treatment of intellectuals and students and was dismissed from his post in January 1987, following similar though far smaller demonstrations for democracy. Now, in mid-May, it seemed to the demonstrators in Tiananmen Square too late for an aging party leadership to arrest the victory of youth over age, of revolution and reform over repression and ideological purity. They were wrong. May 17 marked the floodtide of the revolution. Within three weeks the army was called in to suppress the students and workers in a fusillade of bullets, even though it might cost China dearly in international prestige and economic travail.[1]

What was happening was a shift in Chinese history reminiscent of the great shifts of the past that had occurred time and again when the centralizing power of the state reasserted itself to curb the centrifugal forces that threatened to pull the nation apart. The Zhou dynasty in the fifth century B.C. had thrown up a monarch who, in the words of one observer, had little more power than a "chessboard king." But his dynasty was succeeded two centuries later by the great and cruel centralizing emperor of the Qin dynasty who united the vast country behind him and gave it its name—China.

The assertion of power by Deng Xiaoping was the unyielding reflex of that generation of communist leaders who had weathered the Long March under Mao Zedong in the mid-1930s, led their country to the day of liberation fifteen years later, and survived the chaos of the Cultural Revolution. "We do not fear spilling blood," Deng said a month before the army crushed the demonstrators in Beijing, "and we do not fear the international reaction."[2] The younger generation, which was generally supportive of the students in their demands for democratization and which had ironically shared in the bounty of Deng's economic reforms, was soon destined to take the levers of power into their own hands. In so doing, China might very well shed its image as an exceptional state—the so-called Middle Kingdom that had once demanded the tribute of foreign powers—and learn to live as an ordinary country with institutions that would permit it to be governed by ordinary leaders like a George Bush or a Helmut Kohl.

For China, true stability will only come when it is governed by men and women of less than heroic stature but who understand the need to balance liberty with equality and justice with order. Before the Beijing massacre, China was moving in this direction and, as the old generation passes, may do so again. In this respect, it may well have the support of the army, which found itself divided over the orders it received to quell the demonstrators no matter what the means. For China to grow in material as well as moral strength, the rule of Deng and his cohorts—dictatorial, Stalinist, prisoners of their authoritarian past—has to give way to leaders who understand the need for institutions strong enough to govern a billion people but flexible enough to permit China to be ruled by ordinary men.

The China that was emerging during that Beijing spring was a country whose centralized governing and political apparatus was being decentralized as a result of the economic reforms that Deng had instituted a decade earlier. The essence of these reforms was to dismantle the People's Communes and return the land to individual familes to be farmed privately with long-term leases, to encourage foreign investment, and to allow the growth of individual entrepreneurs who would be responsive to market forces. Farms and factories, for example, could keep for private sale whatever they made after fulfilling the state quota. Deng's aim can be summed up in one word: modernization.

The effects of these reforms were remarkable: a growth rate of almost 10 percent over the past ten years.[3] China needed only to maintain the not unreasonable level of 7 percent growth over the next two decades in order to quadruple its industrial and agricultural output by the end of the century, a process that was further being aided by the high rate of savings, running above 30 percent of GNP since 1970.[4] A 1988 U.S. Commission on Integrated Long-Term Strategy even predicted that China could exceed both Japan and the Soviet Union in economic output by the year 2010.

China's foreign trade had already doubled from 1977 to 1985, faster than any other country in the region. Moreover, 60 percent of all foreign investment in China was in Guangdong Province's three special economic zones that had been set up to encourage foreign investment by offering tax breaks. These economic zones were all on the coast, however, and inevitably disparities grew between the booming coastal regions and the underdeveloped interior (though some largely peasant regions in the interior did very well). Moreover, corruption grew rampant as the coastal entrepreneurs offered good jobs to the sons and daughters of the ruling elite.

As long as the state still controlled certain basic raw materials (such as coal and oil) and access to housing and industrial space, bribes became a common way of obtaining them. There was a perpetual struggle between the entrepreneurs and the centralizers who would punish Canton, for example, by cutting down on its coal and petroleum. Yet despite inefficiencies and inequities, China, in the words of one American scholar, seemed to be moving away "from ideological dogmatism toward eclectic pragmatism, from extreme totalitarianism toward liberalized authoritarianism, from a command economy toward 'market socialism' and from autarkic isolationism toward international independence."[5]

Corruption, however, was not the gravest problem. Workers on fixed incomes saw their purchasing power severely reduced by an inflation that, by 1988, was running over 30 percent. The overheated economy needed to be cooled down, and Prime Minister Li Peng insisted early in 1989 that the government impose a measure of centralized control. The Chinese leaders, to paraphrase an old Chinese proverb, seemed to be searching for the stepping stones while crossing the river.

For many in the party leadership, the central question was how to allow China to develop in such a way as to permit the liberalizing political forces to coexist with an economic liberalism that required new laws to protect and encourage business investment as well as more democracy. As an American lawyer who had been dealing with Chinese corporations put it, laws traditionally have been handed down by the ruler. Historically, the Chinese believed it was better to have no laws and a good ruler. Otherwise, since human nature is flawed, people will try to circumvent laws. Traditionally, the emperor answered to no one but history. He could be overthrown, but otherwise he could do whatever was necessary to preserve his power.

Over time, the Chinese legal system became dominated by criminal law, which meant spelling out in excessive detail the nature and degree of punishment. There were also detailed tax laws. In business affairs, however, there was no contract law as we know it. Matters were handled on a personal basis. As China sought to modernize under Deng, the government saw the need for laws that would regulate foreign economic activity, provide businessmen with legal protection, and rely less on "secret laws"—statutes that existed but were not published. In this respect, the leadership appeared to understand that a modern economic system could not exist without respect for the rule of law. Specifically, here was a country that expected to have an economy that would dovetail with the global economy by relying on a substantial influx of foreign investment, but there was no contract law before the Constitution of 1982, and there is still no corporation law and no banking laws.

The dearth of legal institutions meant that China still depended on the strength and wisdom of a ruler, such as Deng Xiaoping. On the other hand, a trend toward becoming a civil society, as we understand the term in the West, would mean a government of laws, not of men. Civil society, as the China scholar John K. Fairbank has pointed out, is separate from the state except as specified by the law. This requires "a government of laws, not of men, and such aspects of pluralism as a separation between Church and State and between political parties and the government proper." The concept of a civil society has been absent in states under Communist party rule, and, in particular, it is not in the Chinese political tradition. Thus, the power struggle in the party that led to the Beijing massacre cast China back deep into its authoritarian past.[6]

China's past has long been characterized by an isolationism that condemned it to backwardness and impotence. In time, this insularity resulted in the dominance of Asia and the Pacific by other powers, especially by Japan. After the communist takeover, however, China, far from seeking isolation, looked outward for new allies. Even when China was turned in on itself during the Cultural Revolution, it was ripe for the overtures that came from President Nixon to establish official diplomatic ties, for it believed itself also threatened by the Soviet Union. If America saw China as a card to be played against an expanding and threatening Russia, China saw America as its own card to contain the Soviet Union and, perhaps, ultimately Japan. As the Soviet Union grew militarily stronger under Leonid Brezhnev, China and America became virtual allies.

But with the death of Mao in 1976 and Brezhnev less than ten years later, the Sino–Soviet rivalry cooled down. Moreover, under the Ford presidency, with the United States reeling from its defeat in Indochina, China began to distance itself from any de facto alliance with the United States. With Gorbachev's ascent to power and his efforts to end the Cold War, the

Sino–Soviet rift was finally mended. No American policymaker would be able to entertain the notion of playing the "China card" again.

The May 1989 summit meeting between Mikhail Gorbachev and Deng Xiaoping should have been the triumph of Chinese diplomacy. As we have seen, Gorbachev had accepted the three conditions that Deng had laid down for better relations: Soviet withdrawal from Cambodia, a settlement of the Sino-Soviet border dispute, and the scheduled withdrawal of Soviet-backed Vietnamese troops from Cambodia. If anything, the Sino-Soviet meeting was expected to codify the notion that both powers would join with the United States and Japan in maintaining the balance of power in Asia and the Pacific. By establishing a rapprochement with the Soviet Union, Beijing seemed about to find itself with a new flexibility, bound to neither superpower, dependent on no one for its security, and less fearful of aggressive designs on its sovereignty by Soviet Russia. Yet it was Gorbachev's visit that brought forth new student demonstrations, almost toppling Deng at the very moment of his greatest diplomatic success.

Ironically, the effects of the Beijing massacre may well be to isolate China once again. Just as Deng was willing to risk the loss of foreign investment, one of the central goals of his later economic reforms, so too is he likely to find China playing a diminished role on the world stage. During the year following the crackdown, China continued an austerity plan that left the economy at a standstill, with virtually no industrial growth. In the 1980s, a 10 percent growth rate was accompanied at the end by a 40 percent inflation rate; high unemployment with low inflation was the price the Chinese leadership was now prepared to pay for stability.[7]

At the time of the massacre, Deng Xiaoping said that China represented so "tasty a morsel of meat" that the world could not afford to ignore it.[8] That is doubtless true up to a point. But the trend toward a cooperative relationship among the four great Asian powers has been hobbled. Moscow's suggestions for reductions of navies in the Pacific Ocean, the reduction of armed forces and conventional arms in Asia, and "confidence-building measures" for the security of the Pacific sea lanes are not likely to be picked up by the Chinese for some time to come. In addition, the ascendancy of the army in China's political life could imply a more guarded relationship with Moscow. Moreover, Chinese convervatives who are now in charge of the political and economic life of the country disparage Gorbachev's efforts at political and economic liberalization. The spectacle of the Soviet Union threatened with virtual disintegration has only reinforced their feelings. "The free fall of the Soviet economy, the agonies of transition in Poland and East Germany, the political disorder in Yugoslavia," as China specialist Andrew Nathan pointed out, "almost anything that occurs in a world of rapid change shows to [the Chinese leaders'] satisfaction the overweaning importance of order."[9]

Perhaps the most far-reaching effect of the conservative crackdown is the central role the army now plays in China's political life, a trend that reverses the military's role over the past decade when the armed forces were reduced in numbers, money, and prestige. It is unclear, however, exactly how unified the army is in its support of the aging leadership. Two of the surviving marshals of the army, Xu Xiangquian and Nie Rongzhen, in the days preceding the killings called for social order but said that the army should not resort to violence. A few days later, 150 active and retired senior commanders declared in a letter to Deng Xiaoping that the army should never be used to spill the people's blood.[10] In the final showdown, the military leaders acted to preserve the unity of the armed forces.

But Deng Xiaoping, Prime Minister Li Peng, and party leader Jiang Zemin govern at the sufferance of the military. Indeed, the army may come to support a moderate transition precisely in order to protect any wholesale splintering of the officer corps. Nor should we rule out the possibility that riding as a tank commander into the pro-democracy remnants of the student-and-worker movement in Tiananmen Square was a young Bonaparte, prepared to take over with his own brand of reform should disorder follow in the wake of economic distress.

The year before the Beijing massacre, Sino-American economic relations reached a new high. Trade between the two countries had increased by 40 percent to exceed $14 billion. While China became America's thirteenth largest trading partner, the United States was China's second largest export market. By the end of 1991, the United States was running a $10.4 billion trade deficit with China.[11] With some $23.5 billion committed to China, the United States was also the best outside investor after Hong Kong.[12]

And yet, despite Chinese preferences for American trade and aid, China's biggest trade partner is Japan. In the ten-year period 1975 to 1985, Sino-Japanese trade quadrupled, making Japan China's primary source of industrial supplies and technology. Private Japanese investment in China is also high. In 1986, for example, there were 183 Sino-Japanese joint ventures totaling $430 million; by 1988, private investment was close to $1 billion for one month alone.[13] Despite these growing ties, relations between the traditionally hostile powers remain uneasy. One Chinese official put it this way: "Japan does not want China to be strong. Unlike the United States, Japan wants to sell us her exports, but she does not invest in China."

The dreaded specter of Japanese nationalism is never absent. For example, when a recent Japanese textbook series was less than explicit in portraying the prewar hegemonic policies of Japan, and particularly its brutal invasion of China, Beijing lodged a formal protest with Tokyo, and the series was re-edited. Time and again, in conversations with Chinese officials, Americans are being warned against the danger of a Japanese military buildup. Along these lines, officials in China's Foreign Ministry have spoken of the need for an explicit United States military presence in the region and the continued importance of the U.S.-Japan mutual security treaty. A group of experts in the Shanghai Institute of International Affairs noted that if the United States withdraws its "nuclear umbrella" from Japan, Japan has the potential to become a dangerous military power. More crudely put, the Chinese suggested that Japan was like a rich man who was paying the Americans a salary to act as policemen.[14]

Beijing still wants to join the modern world, but on its own terms. Those are not the terms the United States can fully accept if they include continued domestic repression. Under these circumstances, Washington would have to ask itself: What is the likely prognosis for China? On the basis of current economic policy, it would be an uneven growth as the population explodes. While unemployment, along with the government budget deficit, rose in the aftermath of the 1989 crackdown, the economy showed itself to be surprisingly resilient. By 1990, China had a trade surplus, and its foreign-exchange reserves stood at $30 billion, its debt service ratio a modest 10 percent. It would seem that the quasi-private sector, which permitted economic decision making at the provincial and local levels, retained a dynamism of its own. By contrast, the state-owned businesses continued to perform dismally. Where private enterprise flourishes, in food and consumer items, China produces and consumes like a much richer country; in the state enterprises it lags.[15] China's economy, unlike the Soviet Union's and India's, is very much open to the outside world. It will continue to be hobbled by state ownership, but Chinese pragmatism may well prepare the way for a new leadership to crush the power of state enterprises.

Along with political repression, the size and growth of the Chinese population remains a continuing obstacle to prosperity. Most estimates put the Chinese population at 1.25 billion in the year 2000. Under the most favorable conditions, China will not stop growing until it reaches a population of 1.5 billion by 2070. Should this come to pass, with some 30 million unemployed by the end of the century, China could find itself in a potentially revolutionary situation.

A more promising scenario would see more economic growth to satisfy the burgeoning population. To achieve this, however, would require the Chinese to continue decentralizing their decision making and to make a final push for a market-driven economy. This would imply greater participation at all levels, the relaxation of the party's heavy hand, and, in its train, the inevitable push for democracy.

Deng Xiaoping's legacy is a heavy one. As the Yale historian Jonathan Spence has written: "By insisting to the last that economic reforms could be completely divorced from the immensely complex social and cultural effects that the reforms brought in their train, Deng, the party elders and the younger politicians in their clique threatened jointly to commit the government again to the 19th-century fallacy that China could join the modern world entirely on its own terms, sacrificing nothing of its prevailing ideological purity. The task was even more hopeless in the late 1980s than it had been in the 1880s."[16] In the wake of the Beijing massacre, the government finds itself unable to reinvigorate the socialist ideology. While the leadership can compel lip service to ideological purity, true ideological fervor, such as Mao achieved in the 1960s, is no longer feasible.

China may receive foreign assistance from Japan, and it will undoubtedly get some aid from the United States, at least through the offices of the international lending agencies in which Washington plays a controlling role. As a reward for China's support for the United States in its confrontation with Iraq, Washington eased its opposition to World Bank loans, and urged renewal of most-favored-nation status. But until Beijing shows some significant improvement in the treatment of its citizens, America's commitment to human rights should prevent it from offering China sizable credits. Nor is it likely that there will be major U.S. private investment in the Chinese economy. Washington wants, above all, to encourage China to play a major role in the global economy, and to this end, America doubtless will press the Chinese to resume their transition from central planning to a market economy.

America cannot force democratic change in China, but it can reward the Chinese for any movement toward democracy. To the extent that China becomes an ordinary country subject to the ordinary constraints of democratic accountability, to that extent will China become a great power in the modern world. As Americans our role as an exemplar of democracy is one of the most powerful aspects of our foreign policy. The replica of the Statue of Liberty that the students erected in Tiananmen Square on the eve of the crackdown bears witness to this. As an active participant in maintaining the balance of power in Asia, we are also there to contain any Japanese military ambitions. We are there to reassure the smaller nations that they have no fear from a regressive, reactionary China. And we are there to cooperate with the Soviet Union as it seeks a more active and a less confrontational role. Nothing would be more injurious to American interests than immobility—or retreat.

11

America's Pacific Rim

AMERICA is endangered most by those who are willing to settle for a merely reactive policy in the Far East, or what the Bush administration in its first 100 days called the "status quo plus." In fact, the status quo—plus or minus—is the opposite of a prudential policy. American interests require a profound rethinking of our security posture in Asia and the Western Pacific.

The American military presence in Asia and the Pacific was based on the realities of the Cold War. We established bases there to counter communist threats from the Soviet Union and China and to ensure that Japan would never again become a threat. Today the Soviet Union, Japan, and even China are working with the United States to preserve the peace. In addition, the Soviet Union has put pressure on its erstwhile allies, Vietnam and North Korea, to curb their military activities. The extensive American military and naval presence in the region can no longer be justified solely on the basis of containing a potentially aggressive Soviet Union.

The search for a new U.S. security posture means, first of all, seeking negotiations over air, naval, and military deployments as well as arms reductions with the Soviet Union, China, and Japan. But U.S. security is also anchored in two countries to which America is closely tied by historical experience and interests—South Korea and the Philippines. At times because of domestic political considerations, at other times for security reasons, Americans cannot remain indifferent to what happens in those countries.

In South Korea, almost four decades after hostilities have ceased between North and South, no one can discount the possibility of military action. Yet South Korea has prospered mightily in the 1980s. It is the seventeenth largest economy in the noncommunist world, with a per capita GNP of $4,000 (compared to $500 only twenty-five years ago). Though the Koreans have operated a fragile economy highly dependent on the successes of a few large conglomerates, the technocrats are beginning to diversify the economic bases of the country.[1] At the same time, prospects for democracy have been improving in South Korea. President Roh Tae Woo was elected in a free and open election in December 1987, the first direct presidential election in sixteen years, and in the aftermath of the 1988 summer Olympic Games at Seoul, the trend toward full democracy has intensified.

There is also a warming trend in relations between the communist North and the capitalist South that may well lead to renewed efforts on the part of both countries to open up a dialogue on ways of reunifying the peninsula. Pyongyang no longer insists on any withdrawal of American forces as a condition for negotiations. North Korea, it appears, is paying the price of its isolation from its economically well-off neighbors, and it is coming under some pressure from its sometime allies, China and Russia, to reduce tensions in the region; an erratic or highly intransigent North Korea would be detrimental to these efforts. It is more and more likely that North Korea will pursue two approaches: reduce military spending and seek an influx of advanced technology that might be facilitated by an opening to the West.

For its part, South Korea faces considerable pressure to expand ties with the North. In 1988, President Roh called for a six-nation conference, made up of the United States, the Soviet Union, China, and Japan, as well as the two Koreas, "to create an international environment more conducive to peace in

Korea and reunification of the peninsula."[2] He also met with Mikhail Gorbachev in San Francisco in the spring of 1990, with the expectation of establishing diplomatic relations between the two countries; later their relations grew closer with Gorbachev's visit to Korea. In October 1990, Beijing and Seoul agreed to exchange trade offices, and planned for the full diplomatic ties that would follow. And the meeting of the prime ministers of North and South Korea in the fall of 1990 brought forth a new atmosphere of tolerance that could lead to further progress in the future.

As long as North Korea's aging despot, Kim Il Sung, remains in power, however, there is little likelihood of a genuine rapprochement between the two states. Kim was genuinely horrified by what happened in Eastern Europe in 1989 and the Soviet Union in 1991. While both sides offer proposals for reunification, it is not likely to come about without a change of regime in the North. The formula proposed by communists of "one nation, two systems" is doomed. Nonetheless, the North may try to break out of its moribund policies that have left it dreadfully poor, especially compared with the booming South.

North Korea's economy is in even worse shape than the Soviet Union's. It has defaulted on its $3 billion of foreign debt. Its per capita GNP is between $900 and $1,000; South Korea's (as noted earlier) is $4,000, and the gap is worsening.[3] For this reason, despite bitter opposition from the hard-liners and despite Beijing's decision to slow its own program of reform, Kim Il Sung has initiated cautious economic reforms.

There is no compelling reason for most U.S. troops to remain much longer on the peninsula. Even the commander of U.S. forces in South Korea has said that he saw no military need to keep American forces there after the mid-1990s if present trends continue; 7,000 troops are already scheduled to be withdrawn by 1993. He based his conclusion on South Korea's plan to modernize its armed forces.[4] In any case, the eventual aim of the South Korean government should be to defend itself. Roh himself has declared that he wanted South Korea to be militarily self-sufficient by the mid-1990s. Washington could further this process by letting a Korean rather than an American head the ground forces of the Korea–United States Combined Forces Command, which includes mostly South Korean units. But the United States need not phase out all of its troops without getting something in return from North Korea. In this respect, Washington should test North Korea on its offer, made in July 1988, linking force reductions in the North and South to a gradual U.S. withdrawal from the peninsula.

In addition to reducing the levels of conventional weaponry, the United States could unilaterally remove any nuclear weapons it may have deployed or stored in South Korea. This may or may not deter the North Koreans from acquiring nuclear weapons, although the Chinese and the Soviets have absolutely refused to cooperate with the North Koreans in this respect. In any case, U.S. nuclear weapons need not be stored in Korea, because they could be quickly available from other sites in the Pacific. Although North Korea, which has already signed the Nuclear Nonproliferation Treaty, is obliged to accept the safeguards of the International Atomic Energy Agency, it has resisted inspection of its nuclear sites, and Washington may want to delay its withdrawal of nuclear weapons until North Korea agrees on an inspection.[5] In the meantime, German unification stands as a beacon for Koreans longing for one nation on both sides of the thirty-eighth parallel.

South Korea, unlike the smaller littoral states of Asia such as Singapore, Malaysia, or even Thailand, is clearly of central concern to any American administration as Washington seeks to establish a long-term balance of power in Northeast Asia. However, South Korea is unlikely to embroil Washington in a foreign policy disaster that could shake U.S. confidence in its role in Asia. In the Philippines, on the other hand, our armed forces might well be directly engaged in a fluid and unpredictable conflict.

"While I have never varied in my feeling that we had to hold the Philippines," Theodore Roosevelt said in 1901, "I have varied very much in my feelings whether we were to be considered fortunate or unfortunate in having to hold them." The United States finally concluded that it could not and should not hold the Philippines as a colony forever, and independence was granted to the archipelago after the Second World War.[6]

Even after granting independence, however, the United States has remained intimately involved in the Philippines—an involvement that has time and again proved to be a mixed blessing. Our Pacific naval strategy has continued to revolve around our splendid deep-water naval base at Subic Bay and the military airbase at Clark field. But our lease on those bases ended in 1991, and there was widespread opposition within the Philippines to its renewal. We have also continued to play a central role in Philippine politics—both through our support for the regime of Ferdinand Marcos and through our belated support for his democratically elected successor, Corazon Aquino. Aquino's government, and thus U.S. policy, has been under challenge, both from the standpoint of Filipino politics and from a persistent communist insurgency.

There were endless arguments from the Filipinos as to why we should leave the bases. But at bottom, the argument came down to one refrain: the U.S. bases were an obtrusive manifestation of Philippine dependence on the United States. Ridding themselves of the bases was seen by most Filipinos as one way— both real *and* symbolic—of ridding themselves of the plague of dependency.

With the end of the Cold War and faced with the legacy of Philippine resentment, Washington concluded that it did not need the bases indefinitely. The eruption of a volcano from Mount Pinatubo in 1991 destroyed the usefulness of Clark Air Base, and the Pentagon decided not to repair it. By July 1991, an agreement was reached to extend the life of the American naval base at Subic Bay for at least another decade. But that did not satisfy the Philippine Senate, and after lengthy negotiations, a compromise was reached that set the American pullout by 1994.[7] The American naval presence in the Philippines will have lasted almost 100 years.

Is Subic Bay vital to U.S. naval strategy in the Pacific? Although, on balance, retaining it would best fit U.S. naval needs for near term, Philippine bases are not necessary to U.S. deployments. There are basically three choices for the United States: to move the facilities to other existing U.S. bases in the Pacific; to relocate to an expanded base structure in Micronesia; or—the least desirable—to find host countries that would permit new bases to be constructed near the South China Sea. With the bases negotiations concluded, the United States will have ample time to evaluate its needs for bases beyond Hawaii and its own territory of Guam. But it would seem unlikely that further basing arrangements will be required, especially if the trend toward Soviet naval reductions continues.

The bases dilemma was only the most visible problem facing the United States in its relations with the Philippines. A guerrilla war continues to threaten the stability of the government of Corazon Aquino and also poses a threat to the U.S. naval presence until 1994 that could easily provoke an American military response. The New People's Army now numbers about 25,000, and while it is not winning the war, it is not losing it either. To equip the armed forces of the Philippines, for example, the 1989 budget was to be increased by 43 percent over the previous year. Not only did this strengthen the influence and power of the military, but it further drained the deeply troubled Philippine economy.

Worst of all, the government of Corazon Aquino, which the United States tardily helped bring into power in 1986 and has since sustained against threatened coups, failed dismally to put into place the economic and political reforms that Aquino's commitment to democracy promised. After creating democratic institutions, she never put forward a coherent legislative program. The 1986 revolution that toppled Ferdinand Marcos never broke the domination of the country by the same oligarchs and big business people who had always controlled the levers of national power. Moreover, there has never been a social revolution in the history of the Philippines, a country now burdened by heavy foreign debts and the growth of a society of beggars and criminals. A former presidential aide described it this way: "At the heart of it all is the fact that Philippine society is sick, that it is devoid of the values that can make a nation grow and prosper, that it lacks leadership that enables the citizenry to look forward and not to look back."[8] Paradoxically, by trying to avoid authoritarianism, Aquino may be laying the groundwork for a military takeover.

In only three decades, a country that was probably the richest in Asia after Japan has become as poor as Indonesia. Neither democracy nor economic progress has been served by the mounting costs of poor planning, a landless peasantry, and a rich oligarchy. In a country of some 56 million, roughly 60 percent of the population live below the poverty line, which is officially set at about $120 a month for a family of six. While the economy has flourished for the few, the World Bank has singled out the Philippines as having "one of the most unequal income distributions among middle income countries." The Bank estimates that the top 20 percent of the population control over 50 percent of the total income of the country.[9] Corruption (so-called crony capitalism), an underdeveloped agricultural sector, an agrarian reform that never got off the ground because of the determined resistence of an entrenched landowning class who control the powerful congressional committees, profound inequality, and a crushing debt burden—all contribute to the economic disaster that nurtures the insurgency.

While there is talk of a new "Marshall Plan" for the Philippines that could total $5 billion in U.S. aid and contributions by Japan, other Asian nations, and Western European countries, economic aid is no solution. There is little assurance that it would not simply find its way into external bank accounts and prove a windfall for the corrupt "crony capitalists." Instead, a substantial portion of the external debt should probably be written off by the lenders; a genuine land reform program should be gotten under way; and the infrastructure of the country should be upgraded.

Moreover, while land ownership remains a pressing social problem, sheer population growth has led to a shift in the labor force from an agricultural to an urban-based economy. Economic development in the Philippines, if it is to succeed, will require the business elites, with little government interference, to forgo consumption in favor of further investment and to make substantial expenditures for the welfare of their employees.[10]

Can the United States cure corruption and make the Philippines safe for democracy? Surely it cannot. But neither can we ignore the fate of the islands that Americans conquered and ruled for half a century. Under these circumstances, Washington needs to push the Aquino adminstration to fulfill the promises of land reform and to eliminate corruption from the highest levels of government. Though this may well prove an impossible task, the American position on these issues must be absolutely clear and without equivocation. In both the immediate and the long term, such a policy has to lie at the very heart of our relationship with the Filipinos.

Whatever the final outcome of our military needs in the Philippines and South Korea, the aim is not to remove the American presence from Asia and the Pacific. Until some distant date when the two Koreas compose their differences, even if U.S. ground troops are no longer stationed on the peninsula, the United States should guarantee the independence of the South. The U.S. defense commitment to Japan remains important primarily because of the nuclear umbrella America provides; again, this commitment would not be diminished by withdrawing many of the U.S. forces stationed in Japan. The goal of any new U.S. strategic thinking in the Far East, however, should not be to seek a buildup of Japanese military capabilities, even under the guise of burden-sharing, but rather the reduction of them.

Reducing the American military presence should be done with an eye to bringing into negotiations *all* the great powers of the region. It means not only urging the Japanese to reduce their military buildup but also getting the Chinese to curb their nuclear ambitions. Although Washington can initially take unilateral steps to accomplish these ends, it will eventually have to coordinate its efforts with those of the Soviet Union, whose interests coincide with America's in both these respects.

Nor can Washington focus solely on the military aspects of the Asian balance of power. It needs to encourage the Chinese to join international economic organizations with all the responsibilities that that entails, just as it must give Japan a greater say in the operation of these organizations. We cannot expect the Japanese to give more and more foreign aid without according them a commensurate role in institutions such as the World Bank and the International Monetary Fund, which are involved in disbursing these funds.

Most important of all, the United States has to recognize that its security depends on its economic strength—a notion the American public seems about ready to accept. At the end of the 1980s, 56 percent of Americans believed economic competitors like Japan posed a greater threat to America's national security than military rivals like the Soviet Union.[11] This means, in the final analysis, not only bringing about fiscal discipline in order to reduce the size of the U.S. federal deficit and providing incentives for Americans to save more; it means establishing new international economic institutions to ensure currency stability and to promote free trade. Especially in Asia, where Russia, Vietnam, and even China are trying to become integrated in the world economy, where Japan is now the world's leading creditor nation and whose overseas assets will soon exceed $1 trillion, the

United States should be the leader in restructuring the world economy, just as it had to be the leader in restructuring the security perimeter it set up at the end of the Second World War.

The Asian balance of power is not a given. We may not be able to impose a global Pax Americana, but we can promote a Pax Pacifica. Above all, we must recognize and define our strategic priorities—solvency at home, a dynamic policy to maintain the balance of power in Asia, a new security system in Europe, the quest for stability and democracy in Mexico and the Caribbean, and the remaking of the international economic order. These are the new but hardly inconsiderable requirements of American power at the end of the century.

PART IV

The Hemispheric Commitment

12

The Legacy of Mexico

THE necessity for a balance of power in Asia and the reality of a Europe without frontiers may be more obvious than the third leg in the new triad of American foreign policy—the hemispheric commitment. The United States and Canada have been welded even more tightly together through the free-trade area established in 1988. Canada, a founding member of NATO, is already America's largest trading partner. Canadian nationalists, always wary of the threatening image of a domineering America, will doubtless seek to protect Canada's heritage from American cultural domination. But in a relatively prosperous world, U.S.-Canadian relations are also likely to prosper; even in more straitened economic circumstances, the likelihood is that the United States and Canada, despite the separatist yearnings of Quebec, will cling together against all others.

More problematical, and hence more worrisome, are U.S.-Mexican relations, and indeed Washington's central connection to Latin America. Yet, looking to the twenty-first century, the likelihood is that the United States will have to direct the bulk of its energies to the northern half of its own hemisphere, and especially to Mexico, Central America, and the Caribbean Basin.

The legacy of U.S. involvement in that region has been a particularly unhappy one. In 1848, America engaged in a war that deprived the Mexicans of one-third of their territory. James Polk's Mexican War was not fought to secure the nation against a foreign threat, but to assert American predominance on the continent. Sixty-five years later, Woodrow Wilson rejected all notions of Manifest Destiny and expansionism, declaring in the first year of his presidency that "the United States will never again seek one additional foot of territory by conquest."[1] In Wilson's view, trade had superseded annexation as the primary U.S. concern. For if the profits of trade could be had, and its agents protected without the trouble of actual administration, then annexationism had indeed become an archaic concept. Nonetheless, Wilson sent troops into Mexico, Haiti, the Dominican Republic, and Nicaragua. (In fact, Theodore Roosevelt and Wilson sent U.S. marines and sailors into Central America at least fifteen times between 1902 and 1920.) In Mexico, this was done to establish the principle of good government, without which, in Wilson's view, anarchy would prevail. This in turn would spell financial loss for Americans in Mexico, as well as providing an almost irresistible temptation to foreign powers to increase their influence south of the border. ("I am going to teach the South American republics to elect good men," Wilson declared in typically didactic fashion.)

Toward the other countries of the region, Wilson may have abandoned the philosophically expansionist rhetoric of a Thomas Jefferson, but he had replaced it by the commercial rhetoric that declared that trade was the method by which the civilizing influence of the United States was now to be exported. Committed to avoid annexation, Wilson endorsed a policy that was certainly strengthened by his acceptance of interventionism. In the years to come, that behavior was to become habitual. It was officially expressed in January 1927, by Undersecretary of State Robert Olds: "The Central American area down to and including the isthmus of Panama constitutes a legitimate sphere of influence for the United States, if we are to have due regard for our own safety and protection. . . . We do control the destinies of Central America and we do so for the simple reason that the national interest absolutely dictates such a course. There is no room for outside influence other than ours in the region."[2]

Indeed, it was not until Franklin D. Roosevelt proclaimed a new policy of "the good neighbor" that the United States curbed its habit of intervening directly in the Caribbean, Mexico, and Central America. During World War II, U.S. and Latin American military officers cooperated closely to prevent any intrusion into the hemisphere by German or Japanese forces. By 1945, the United States had become the preeminent supplier of military equipment and instruction for Latin American armies, whose efforts (coinciding with Washington's own preferences) were centered as much on maintaining internal stability as they were on repelling any Axis aggression. The United States, especially during the early stages of global conflict, badly needed stability and support in its own hemisphere. In helping Washington pursue this goal, the leaders of Central America never wavered.

But by the end of the war, the U.S. government had come to believe that the danger threatening the hemisphere was "Communist political aggression against the hemisphere," as a State Department official put it.[3] The fear of communism as the key determinant of U.S. policy marks a serious turning point. Much less was now heard about the need for markets, though North American companies were obviously still important and their interests in Central America, the Caribbean, and even Mexico were close to those of the State Department. The key objective of U.S. policy, however, was to keep the five states of Central America—and, if possible, all of Latin America—from going communist.

Throughout the years of the Cold War, the United States constantly renewed its reliance on military elites and authoritarian rulers. As a consequence, the agents of social change in the region were forced to seek help elsewhere, most notably from Cuba after Fidel Castro took power in 1959, when he sought the Soviet Union as his ally and protector. As Washington saw it, democracy (or at least a more equitable social and political structure) was desirable in Central America, but the countries themselves were untrustworthy for the task.[4]

The loss of Cuba to a Marxist regime, dreaded as a vital blow to U.S. security, in fact proved easily manageable, once American interests were clearly spelled out after the 1962 missile crisis. The threats to the Panama Canal and Caribbean sea lanes that had often been predicted as consequences of a hostile takeover in Havana did not materialize. But Latin American suspicions about U.S. motives and methods were greatly heightened by the crude and incompetent anticommunist actions undertaken by Washington after Castro's victory. Similarly, the "loss" of Nicaragua to the Marxist Sandinistas barely affected American security, contrary to the fanatic ideologizing of conservatives in the Reagan administration. Managua's swing to Marxism proved less of a regional irritant than did the contra war supported by Reagan and funded to a large extent by various shadowy private backers.

In truth, the Latin American crisis of the Cold War era had far more to do with domestic social stratification and economic mismanagement than it did with the spread of Marxism. By the time of the American invasion of Grenada in 1983, many key Latin American countries—most notably Mexico, Brazil, and Argentina—had sunk so deeply into debt to foreign governments and banks that there seemed little likelihood of their ever making good on the loans. Much of the income of these nations was needed to pay off the interest on their debts. Wealthy Latin Americans, especially Mexicans, took funds out of their unstable national economies and deposited them abroad, creating staggering liquidity problems. At the same time, their poorer countrymen fled in growing numbers from domestic violence and unemployment to the sanctuary of the United States. A hemispheric crisis was no longer imminent; it had already arrived with devastating force.[5]

No matter how the national interest is defined, no one would deny that a crisis in Mexico would centrally affect the United States. In this respect, any U.S. policy that looks beyond the Cold War must focus on political and economic stability south of the Rio Grande. Mexico, however, has been ruled by one political party for more than half a century, its army plays no role in its politics, and its foreign policy has largely been directed at protecting Mexico from the hegemonic ambitions of its great neighbor to the north. But it is nonetheless difficult for Washington not to view Mexico through the narrow lens of America's concern with stability. Mexico, burdened with a population that is increasing at such a pace that it is expected to reach 100 million by the end of the century, is by far the greatest source of migration, legal and illegal, to the United States.[6]

Its foreign debt by 1989 stood near $100 million, and in its effort to cope with the service charges, the Mexican government was forced to invoke austerity policies to reduce public spending and the state's role in the economy.

But the effects of austerity only exacerbated Mexico's problems. For example, in 1982, Mexico was spending 5.5 percent of its gross domestic product on education; by 1989, it was spending less than 3 percent. In 1982, Mexico was was spending 2.5 percent of GDP on health; in 1987, it was spending less than 1.5 percent. In the early 1980s, Mexico had one of the most advanced infrastructures in the so-called Third World. Today Mexico's phones may be the worst in Latin America, its highways are dilapidated, its electrical grid run down, and its oil installations poorly maintained. According to the Mexican political scientist Jorge Casteñeda, in the last decade of this century Mexico must create "a million jobs a year just to accommodate the new entries into the labor market. For that to happen the economy has to grow at 4 to 5 percent a year."[7] If this does not occur—and it is unlikely to do so—millions of Mexicans will have two stark choices: being unemployed on the streets or entering the United States to find work.

What happened to Mexico, a land rich in opportunity and promise thirty years ago but a crisis-ridden, often pauperized society by the 1990s? Throughout the postwar years and well into the 1960s, Mexico, by following a state policy of import substitution, enjoyed an unprecedented spate of growth stimulated by the development of a considerable industrial sector. This growth strategy also served to promote steady price stability and provided jobs for an increasing population. All seemed serene, as Mexico's gross national product expanded at an average rate of 6 percent from 1940 to 1965, accompanied by only a moderate amount of borrowing, which in itself contributed to the expansion.[8]

Mexico's essentially inward-looking policies would cause problems for the country only later, since they tended to isolate the economy from international pressures. As industrialization proceeded, Mexican industry had to import more and more spare parts, while the possibilities for expanding production of consumer goods were almost exhausted even as early as the 1950s.[9] At the same time, agriculture began to stagnate; government policies were so focused on manufacturing that the agricultural sector of the economy faltered. Small farmers received little investment or assistance, and this in turn served to widen the already alarming income gap between the rich and the poor.[10]

The growing poverty seemed to require a shift in strategy, and in 1970 the new president, Luis Echeverria, decided to attack the income gap through more public expenditure. In the six years of his administration, government-owned enterprises rose from 86 in number to 780. As spending grew, so did the bureaucracy: government employees nearly doubled.[11] It was a policy fraught with deadly consequences. Unable to produce a surplus in foreign exchange, the government turned to international lenders for financing. As a result, foreign debt owned by the public sector nearly quintupled.[12] Rampant inflation followed, accompanied by a wave of capital flight.

José López Portillo's term as president began in 1977 with the discovery of immense new oil reserves. Given the sharp rise in the cost of oil that occurred during the 1970s, Mexico's confidence that it could handle its growing problems grew to new heights. By the end of the decade, the administration had significantly increased its expenditures and embarked on a program of ambitious new investment projects. Oil extraction rose rapidly, and by 1978 the economy was once again booming along at a growth rate of 8.3 percent.[13] The share of crude oil as a percentage of total exports rose from zero in 1973 to 69 percent in 1981.[14] Anticipating future extraction at oil's 1979 prices, government expenditures soared. López Portillo unwisely turned to foreign sources to make up the difference.

The infamous debt crisis of August 1982 was the product of a number of converging problems, not the least of which was the administration's unprecedented level of borrowing. And commercial banks were only too eager to lend. Falling world oil prices led to more borrowing. Foreign debt reached nearly $100 billion by 1983. Meanwhile, interest rates had begun to climb, which put more pressure on Mexico to come up with more revenue. Recession in the industrialized world reduced demand for Mexican non-oil exports, while imports tripled (in real terms) between 1979 and 1981.[15] With reserves depleted and unable to service its debt, Mexico plunged into virtual bankruptcy by 1982.[16]

Beginning with the August 1982 meeting between U.S. representatives and the Mexican finance minister, which resulted in a renegotiation of the debt, Mexico was forced to embark on an austerity program designed by the International Monetary Fund to curtail public spending. Subsidies for goods and services were slashed, including for such basic items as tortillas, beans, and water. Domestic industrial production fell sharply, not only because of reduced consumer purchasing power but also because foreign exchange was so limited that imported goods often could not be obtained.

To cope with the crisis he inherited, López Portillo's successor, Miguel de la Madrid Hurtado, presided over a bitter austerity program from 1982 to 1988 that cost the average worker 40 to 50 percent of his purchasing power; the peso underwent a near 100 percent devaluation.[17] To restore the country to solvency, de la Madrid broke with the isolationist policies of the past, especially with the traditionally high degree of state intervention. He was singularly successful in this effort: the financial deficit of the public sector, for example, decreased from 17.6 percent of GNP in 1983 to 9.9 percent in 1985. It was, as a highly placed Mexican bureaucrat put it, "an adjustment without parallel in the world."[18]

Despite de la Madrid's faithful adherence to the canons of the international financial community, the commercial banks were unwilling to expose themselves further. As a result, Mexico found itself hard put to get new loans except through international lending agencies. Since the 1982 financial crisis, there have been numerous attempts to solve the debt crisis, both in Mexico and in Latin America as a whole: adjustment loans, debt-equity swaps, discounting the debt—but no forgiveness. If the United States, for example, were to forgive the Mexican debt, nations such as Brazil and Argentina would naturally seek a similar arrangement. And since it would not be economically feasible for our country to forgive all debts, we would have to devise criteria to determine which, if any, additional debts would be forgiven and which would not.

But criteria based on what? Importance to our national security, or to our business community? Willingness to assist in achieving U.S. goals in the hemisphere? Record on human rights? Under this scenario, forgiving debts would come dangerously close to a punishment-and-reward system, which is never a sound basis for the conduct of foreign relations. Because of this, and because of the effects it would have on the U.S. budget, forgiving the debt is neither a realistic nor a desirable policy.

De la Madrid's successor in 1989, Carlos Salinas de Gortari, was greeted with a new American plan to relieve the burden of Mexico's debt. The so-called Brady Plan, written by U.S. Secretary of the Treasury Nicholas F. Brady, gave banks new opportunities to reduce their outstanding loans, in return for which they were expected to lend new money. While the banks were only too willing to reduce their loans through swapping debt for low-interest bonds, they were reluctant to extend new credit. In short, the banks reduced their claims on Mexico and increased their reserve funds, but they proved singularly unwilling to risk their capital on new credits to Third World countries. By the end of the 1980s there was no banking crisis, but there was a continued crisis of development in Mexico and the other debt-ridden Latin American countries.[19]

Mexico's only long-term hope of digging itself out of the quagmire of debt was to increase its exports and attract new investment. To attain these goals, Mexico's policies in the 1990s are focusing on dismantling import licensing and tariff barriers to trade. In 1982, for example, 100 percent of the value of Mexican imports was subject to prior licensing; six years later, only 20 percent remained subject to control. By the end of de la Madrid's term, these efforts to liberalize Mexican markets were rewarded with a shift in the nation's trade balance from deficit to surplus. The economy was on the move again. Salinas, in turn, continued to push the privatization of the public sector, including proposals to return control of the banking system to private hands; over 70 percent of the state's corporations were privatized by 1990. Moreover, by enforcing the tax laws while cutting expenditures, Salinas increased revenues by over 13 percent, so that the fiscal deficit shrank from almost 12 percent in 1987 to less than 5 percent by 1990. Perhaps of even greater significance, Mexico also adhered to the General Agreement on Tariffs and Trade (GATT), which signaled its long-term commitment to a more open market. One of the world's most protected economies was rapidly becoming one of the world's most open.[20]

The most important development along these lines, however, came in the spring of 1990, when Salinas and George Bush said they would seek to negotiate a bilateral trade agree-

ment that would lead to a continent-wide North American Free
Trade Zone—an audacious move by a Mexican president. Mexico is already America's third largest trading partner after
Canada and Japan. The United States is Mexico's primary trading partner, with Japan second.[21] There is marked asymmetry in
the United States–Mexico trading patterns, however, for
whereas Mexico's trade with the United States amounts to
almost two-thirds of its total, U.S. trade with Mexico is only 5 to
7 percent. Moreover, the volume of U.S. trade with Mexico is
small compared with U.S. trade with the other industrialized
nations of the world. Latin America as a whole, for example,
accounts for less than one percent of manufactured goods
imported into the United States; Mexico, on the other hand,
relies on the United States for the export of 80 percent of its
manufactured goods.[22]

Mexico has indeed taken numerous measures to increase
the volume of private foreign investment. In 1989, new regulations automatically granted 100 percent foreign ownership to
all new investments valued under $100 million; previously, for-
eign investment was discouraged by limiting foreign ownership
to 49 percent and requiring advance approval from a National
Investment Commission. Nonetheless, foreign investment still
amounts to less than 5 percent of Mexico's gross domestic
product.

In the 1980s, the so-called *maquiladoras* represented the
greatest portion of foreign investment. These are foreign-
owned and -operated assembly plants located near the
U.S.–Mexican border, which are designed to take maximum
advantage of cheap Mexican labor to assemble imported components for re-export. No less than 98 percent of all material
inputs are imported from the United States, Japan, and
Europe.[23] By the end of the 1980s, over 1,000 assembly plants
employed over 279,000 workers. They yielded $1.5 billion in
earnings, making the *maquiladora* program Mexico's third
largest single source of hard currency.[24]

Despite these figures, the gains rest largely in the hands of
the foreigners. Since the companies are owned by them, profits
are remitted abroad, and the taxes earned off them are small
compared with the profits earned by the foreign owners. Moreover, as Jorge Casteñeda has written, "Since the physical installations are cheap and easy to build and remove, any shift in the

American business cycle can bring sudden and dramatic changes in the plant economy. Many firms simply pack up and leave, literally overnight, as they have been quitting Asia for Mexico in recent years." In Casteñeda's view, echoing that of many thoughtful Mexicans, "instead of an outward-looking sector of the Mexican economy, the *maquiladoras* seem to be an extension of the American economy on Mexican soil. A welcome extension because of the badly needed dollars it brings, but a reluctantly accepted one because of its contribution to the rampant integration of two unequally matched economies."[25]

The need for direct U.S. investment was underscored when Salinas discovered that little investment capital was available for the Western Hemisphere from Western Europe after the collapse of communist rule in Eastern Europe. As Salinas said at the time of his meeting with Bush in June 1990, "The changes in Europe and East Asia and an apparent reliance on blocs convinced me that we should also try to be part of an economic trading bloc with the United States and Canada. But," he insisted, "we do not want this bloc to be a fortress."[26] This view was apparently fully shared by Bush.

Despite the doubts expressed by many Mexicans over the desirability of integrating the Mexican with the American economy, it seems likely that such integration cannot be avoided. The arguments for a North American Free Trade Area are persuasive: it would greatly reduce the risk of disruption for businessmen on both sides of the Rio Grande,[27] and a formal agreement would benefit Mexico because it would provide rules that could increase Mexico's control of its economic destiny. Canada will undoubtedly join the free-trade zone, which would bind the three countries' economies into a common market that would include more than 350 million people, larger than the European Community. In 1990, Bush also proposed, in what he called his "Enterprise for the Americas" initiative, a vaguely formulated scheme for a free trade area stretching from the Canadian glaciers to Tierra del Fuego.

While Mexican businessmen support free trade in principle, they show real hesitation when it comes to putting their ideals into practice. Most of their business is not labor-intensive but capital-intensive, and they are likely to be badly hurt unless the free-trade agreement is phased in slowly. Many heavy industries are trying to convert to service industries, and Mexico does not want free trade in the financial services. Nonetheless, while there is genuine fear of economic domination by the United States, Mexican nationalism no longer holds sway, even among the opponents of the Salinas government.

Nor is there any hard evidence that free trade will lead to greater democracy, though privatization and decentralization of the economy may well help decentralize entrenched political power.[28] Salinas' party, the Institutional Revolutionary party (or PRI), almost certainly stole the 1988 presidential election. And while Salinas has opened up the economy, he has been reluctant to open up the political system. In Salinas' thinking, free elections should be postponed until the economy is rolling and wages have risen, and, under these circumstances, people will return to their traditional support of the PRI. Robert Pastor, formerly of the National Security Council and now a professor at Emory University, makes the case for integration this way: "If Mexico does not succeed in modernizing its economy and democratizing its politics, the United States cannot escape the consequences. Instability in Mexico would cause massive migration, capital flight, and radicalism."[29] This is hardly in the U.S. national interest.

There is a heavy irony in this debate over the desirability of integration. In Pastor's construct, the United States now presses for integration because Americans believe they are a nation of problem solvers. But if the United States finally overcomes Mexico's fears of being swallowed up by the economic leviathan to the north, within a few years the two nations might exchange roles. American labor unions might find themselves complaining over an increase in unemployment and the flood of cheap Mexican goods entering the U.S. market; Mexico, on the other hand, would insist on more barriers going down and that the United States "not succumb to interest-group pressures and short-term anxieties."[30] But perhaps "debate" is the wrong word as we look at the future of U.S.–Mexican relations. It is no longer a question of whether integration of the two economies is desirable, but rather how and when it will take place. In that case, in the hierarchy of U.S. vital interests, Mexico stands at the very summit.

13

Inescapable Entanglements

For much of the 1980s, events in Central America bedeviled and obsessed the Reagan and Bush administrations. First came the success of the Nicaraguan revolution in 1979 that brought the Sandinistas to power; soon after, a conflict in El Salvador between the Marxist guerrillas and a military-dominated government erupted into what became a seemingly endless war. Despite the Reagan administration's funding of the anti-Sandinista contras, despite over $4 billion in military and economic aid poured into El Salvador, Reagan succeeded neither in ridding Nicaragua of the Sandinistas nor in decisively denying territory to the Salvadoran rebels, the Farabundo Martí National Liberation Front (FMLN).

Then, in the first year of the Bush administration, the Salvadoran guerrillas launched an offensive in November 1989 that demonstrated their staying power and the Salvadoran army's continued inability to come up with a convincing scenario for victory. While atrocities were committed on both sides, the army's wanton slaughter of six Jesuit priests during the November uprising underlined the problem the United States faced in curbing the human rights violations of an ally.

The lack of a courageous and independent Salvadoran judiciary seemed especially to mock America's effort to democratize the country and professionalize the army. Nonetheless, there was real hope for a peace in the region. Violeta Barrios de Chamorro, who headed a fourteen-party coalition, was elected president of Nicaragua in a stunning upset of the ruling Sandinistas. This meant the end of America's aid to a rebel army—the contras—and a shift in policy toward El Salvador that stressed negotations as the only way to settle the war.[1] The Caribbean-Central American sphere, like it nor not, remains a U.S. sphere of influence, perceived as such by its inhabitants. This requires the United States to support democratic norms and economic development; but, above all, it means that the United States should work for the demilitarization of the region.

In Nicaragua, it was not Washington but the peace plan drafted by Costa Rican President Oscar Arias Sánchez in August 1987 that provided a mechanism for negotiation among the forces contending for power in Nicaragua—the armed contras, the internal opposition political parties, and the ruling Sandinistas. Signed by the five presidents of the Central American nations, the Arias plan dealt with the need for democratic reform in the region rather than with external security. Although the plan was supposed to apply to all countries in the region, it was designed primarily to promote a settlement in Nicaragua. Moreover, to everyone's astonishment it succeeded in this respect, for it helped to bring about the disbanding of the U.S.-backed contra organization and paved the way for nationwide elections in February 1990 for a new president of Nicaragua, elections that were monitored by international organizations. It also led to the resumed contacts between the Salvadoran government and the guerrillas, and to the decisions of the political leaders of the rebel front to return to El Salvador.

The Arias plan may have been decisive in propelling the presidents of the five Central American nations to negotiate settlements without the direct participation of the United States; but further pressure for change came from the Soviet Union. The reformist leadership of Mikhail Gorbachev made it clear to the Sandinistas that their centralized economic program had to be modified if they were to expect continued economic aid from Moscow; moreover, the Soviets had no intention of supporting the Nicaraguans in any military confrontation with the United States. Asked how the Soviet Union would respond to an armed American attack against Nicaragua, Sandinista ideologue Bayardo Arce replied: "The only response would be a protest."[2]

The Sandinista revolution, once seen as the vanguard of revolutionary change throughout Central America, was spent by the end of the 1980s. Nicaragua had not become a Soviet military base, as many American conservatives had suggested. Between 1981 and 1988, according to a group of international experts called in by Sandinista President Daniel Ortega to evaluate the economy, consumption had been cut by 70 percent. Real wages of Nicaraguan workers had fallen to less than 10 percent of their level in 1981, and per capita GDP had dropped to roughly $300 per year, even less than that of Haiti, long considered the poorest country in the Western Hemisphere.[3]

Certainly the U.S. trade embargo, imposed in May 1985, and the costs of fighting the U.S.-backed contras, put a severe burden on the economy, with defense expenditures consuming about half the budget, or about 17 percent of the gross domestic product. In addition, Washington had forced a cutoff of access to loans from the International Monetary Fund, which further deprived Nicaragua of liquid assets and short-term credits. But the Sandinistas have to bear much of the blame for the mismanagement of the economy. After the 1979 revolution, about 40 percent of cultivated land was confiscated and turned over to large-scale cooperatives or state farms, though a small amount was later redistributed to individual farmers. Nevertheless, even under the Sandinistas, over one-half of the economy remained in private hands. But the banking system was run by the state, and it controlled not only all wages and prices but also import licenses and decisions on where to sell and to whom. This system of central planning proved disastrous. Even during the mid-1980s, when the contra war was at its most intense, the Soviet Union and East Germany, both of which then provided sizable amounts of military and economic aid, told the Sandinistas that they "could do much more with the resources" they had.[4]

If nothing else, Violeta de Chamorro's victory revealed the fierce discontent that the population felt for the Sandinistas after years of feeling the effects of their rigid and futile economic policies. Although both the victors and the vanquished in the Nicaraguan elections showed a remarkable degree of pragmatism and a genuine desire for reconciliation in the immediate aftermath of Chamorro's triumph, the likelihood for Nicaragua is continued factional strife. The Sandinistas remain the strongest single party in the National Assembly, holding 40 percent of the seats there. Chamorro's coalition, UNO, spans too wide a spectrum of ideologies to hold together in the long run. The history of Nicaraguan politics, moreover, is the story of factionalism gone wild.

A factionalized Nicaragua, however, is unlikely to command the attention of the United States. Even when Reagan was most loudly proclaiming the dangers to U.S. security of a Sandinista government supported by the Soviet bloc, he was unable to persuade the American people to support his policy of arming counterrevolutionary forces to topple the Nicaraguan regime. Nevertheless, it is also hard to imagine that the United States will abandon its presumption that any Nicaraguan government must be responsive to U.S. needs as defined by Washington.

Whatever Moscow's intentions in the early 1980s, the Soviet Union today is hardly likely to challenge the United States so directly in America's sphere of influence. Gorbachev surely spoke the epitaph for a Soviet policy that would threaten the United States in Central America when he chose Havana as the place in 1989 to declare: "We are against doctrines that endorse the export of revolution or counterrevolution."[5] A year later, Moscow joined with Washington to urge the Salvadoran government and the leftist guerrillas to reach a ceasefire and a political compromise.[6]

The policy for the United States at this moment in history should be to work for the demilitarization of the region. In the first instance this means reducing America's military presence in Honduras, which in the 1980s became a virtual aircraft carrier for our support of the contra army and for intimidation of small countries that would deny our will—what might be called "gunship diplomacy," in which nations themselves were turned into gunships. It means urging negotiations between the Guatemalan government, which is effectively controlled by a repressive military, and its principal leftist opponents. Above all, it means keeping the pressure on for a negotiated solution to the fratricidal war in El Salvador.

El Salvador, far more than Nicaragua, has posed the most deadly challenge for American diplomacy. For over a decade there has been no government in power that could control the army. The army, in turn, has countenanced grave abuses of human rights by allowing the so-called death squads, backed by the military, to assassinate real or supposed activists on the left. Years of U.S. economic and military assistance to El Salvador seems to have made little difference in the outcome of the war. Even the former head of the U.S. Southern Command, General Maxwell R. Thurmond, told Congress that the Salvadoran army could not defeat the rebels and that the only way to end the fighting was through negotiations.[7] In 1991, as they had for over six years, the guerrillas controlled about one-third of the country. The army, better trained and equipped than it was ten years earlier, had made no significant headway in winning the war. But the rebels, too, could not win a military victory. They could "win" only by not losing.

Both the Reagan and Bush administrations were convinced that the best way to prosecute the war was to "professionalize" the Salvadoran military. This was not to become a war fought by American soldiers, but Washington would, and did, re-equip, expand, and retrain the Salvadoran army. In 1983, a National Campaign Plan was even devised by the U.S. military and accepted by the Salvadoran military. This plan was to involve the Salvadoran armed forces in winning the hearts and minds of the people aiding the guerrillas. It meant involving the army in civil action. And it did not work.

In a devastating critique of the war, four field-grade U.S. officers wrote in 1988 that after billions of dollars of U.S. economic and military aid, after years of training Salvadoran officers in U.S. camps, El Salvador's dependence on the United States was "near total." By the end of the 1980s, over 50 percent of the Salvadoran government's budget came from the United States. The authors of the study said that those they interviewed "were unanimous in asserting that terminating American support for El Salvador with guerrillas still in the field will result in [the Salvadoran armed forces'] defeat and the collapse of [the government]."[8]

The reasons for this state of affairs were many and varied, but, at bottom, they came down to three points: the use of the Salvadoran military as an engine of political reform, the use of conventional tactics in a guerrilla war, and government corruption. Since at least the mid-1960s it has been characteristic of American policy in Latin America to use the military to try to bring about reform. This was tried in Panama when the Panamanian military under Omar Torrijos was encouraged to engage in civic action projects, such as building schools and hospitals, with apparently little thought given to the corruption that these projects inevitably produced.

In El Salvador, the Americans saw the armed forces as the "closest thing to an effective national institution." Security assistance was, moreover, controlled by the Salvadoran army, not by the U.S. advisers; it was generally spent on high technology and heavy weaponry that are ill suited for a counterinsurgency operation. More and more was spent on helicopters, which meant keeping the men in the air above the people, rather than on the ground where they might seize and hold territory. It was the Vietnamization of the Salvadoran war.

As for the American effort to change the nature of the Salvadoran armed forces, that, too, was doomed. "Their vision stops where it begins to change their lifestyle," one American officer said.[9] The quality of the Salvadoran military leadership is reflected in that remark. Whole classes from the military academy are promoted, irrespective of individual talent and initiative. Salvadoran small-unit tactics have been described by some Americans as "search-and-avoid patrols." Nor are the foot soldiers of the Salvadoran army drawn from the middle and upper classes; this is war fought by the poor, impressed into military service. It is hardly a patriotic struggle.[10]

With the electoral defeat of the Sandinistas—the rebels' ally and main arms supplier—and nervous over the economic crisis in Cuba, the FMLN changed tactics and embraced negotiations. On the government side as well, the 1989 election of President Alfredo Cristiani of the right-wing Arena party had made it possible to achieve a settlement that would probably gain the broad support of the conservatives in the National Assembly and perhaps even of the military. In April 1991, aided by a U.N. mediator, the two sides agreed to provisions establishing civilian control of the military, providing for independent oversight of free elections, and strengthening the judiciary to ensure an investigation and prosecution of human rights abuses. Then, in September 1991, after 75,000 deaths in 12 years of civil war, the Salvadoran rebels and the government reached a broad agreement that promised peace. The guerrillas were to be allowed to seek appointment in a civilian-controlled police force, and the government was prepared to protect the right of guerrilla families to hold on to land they have occupied. The U.N. would play a signal role in seeing to it that these provisions were carried out. Yet both sides remained armed, and it will take not only the efforts of the U.N. mediation team but also of the United States to see to it that free elections and a democratic system based on law come to pass. While it is encouraging that military officers responsible for killing the Jesuit priests were also finally convicted of murder, it will not be an easy task to purge the army of its homicidal members or to install an independent judiciary.[11]

For the United States, the only real leverage we possess, should negotiations break down due to government intransigence, is for Congress to radically cut military aid and cash transfers each year until serious talks to settle the war resume. Aid to police forces must be linked to judicial reform, including the prosecution and punishment of members of the armed forces who violate human rights; and economic aid should be allocated for basic needs, reversing the policies of the Reagan administration that provided three times as much funding for waging war as for development and reform.[12]

What would happen if the United States were to walk away before a successful negotiation? The likely consequences might be either an eventual victory by the guerrillas or a genocidal war waged by the army, as the Guatemalan army conducted in the early 1980s and as the Salvadoran army itself did in 1932. Yet a policy of endless war is no policy at all. Faced with such a situation, there is comparatively little we can do. If the leaders of any nation are willing to use mass murder, extortion, and torture as instruments of national policy, and other countries are unwilling or unable to intervene directly in the affairs of that nation, then few choices are left. We can attempt to apply varying degrees of pressure by cutting aid and reducing our commercial involvement, or we can go the full course and treat such countries as "pariah nations," witholding any and all support and urging our allies to do likewise. There is, of course, no guarantee that this will alter the behavior of the country in question. But *no* policy can supply such a guarantee, except perhaps unilateral U.S. military intervention, which, with the ending of the Cold War, is something the American people will not tolerate.

The American involvement in Panama parallels our interest in the Philippines. While the United States has never occupied Panama, it certainly created it. Without the U.S. gunship off the Atlantic coast in 1903, Panama would have doubtless remained an appendage of Colombia. But Theodore Roosevelt wanted a canal across the isthmus. When the first canal treaty was signed between Secretary of State John Hay and the French engineer Philippe Bunau-Varilla, no Panamanian signature was thought necessary. Later, when Roosevelt sought to find a legal defense for his behavior, his attorney general is said to have remarked: "Oh, Mr. President, do not let so great an achievement suffer from any taint of legality."[13]

Under the terms of the treaty, the United States was granted control "in perpetuity" over the canal and the Canal Zone. Under the Panama Canal treaties signed between Jimmy Carter and General Omar Torrijos in 1977, the canal is to come under Panamanian control on December 31, 1999. Yet the United States is permitted to act *unilaterally* in defense of the canal *forever*. The United States also has the right to station forces and retain bases in Panama until the year 2000. The treaties give Washington no right to interfere in the internal affairs of Panama, and their signing generally defused the anti-American sentiment that had been raging during the 1960s over the American presence.

Yet the Bush administration chose to intervene nakedly in Panamanian internal affairs with the dispatch of over 20,000 troops to the isthmus in December 1989. This had little to do with the Cold War. It was, in essence, an atavistic return to the past, the desperate attempt to salvage the remains of a failed policy that resulted from American efforts to turn the Panamanian National Guard into an army to defend the indefensible—the Panama Canal.

From 1903 to 1968, Panama had a constitutional system of

government, dominated by a commercial oligarchy. During this period, the armed forces were little more than a militarized police force that wore no firearms, though they periodically intervened to back presidents who would not challenge their prerogatives. All this changed in 1968 when the charismatic populist leader Colonel Omar Torrijos Herrera and the National Guard overthrew the legally elected president, the nationalist Arnulfo Arias.[14] Torrijos also portrayed himself as a Panamanian nationalist. The National Guard, which numbered about 6,000 troops when he took power, was made up mostly of blacks or of middle-class mestizos like Torrijos himself, who was the son of a rural schoolteacher. National Guard officers were enthusiastic about *"Torrijismo"*; they stressed programs to improve conditions in rural regions and to promote the poorer classes to political power. During Torrijos' thirteen years as president before he perished in an air crash in 1981, the National Guard built schools and hospitals, repaired roads and bridges, and constructed low-cost housing for the *campesinos* who had come to the cities; and in these activities the army had the moral and economic support of the United States. This was toward the end of the era of the Alliance for Progress, when Washington believed that the armed forces should be encouraged to work among the people, winning "hearts and minds," in order to forestall Marxist uprisings.[15]

When Torrijos died, the country was thrown into political turmoil. Between 1982 and 1985, the National Guard overthrew three powerless presidents while plotting to share power among themselves. By 1984, General Manuel Antonio Noriega had become the head of the military, and with the rigging of the 1984 election (to prevent Arnulfo Arias from once again being elected), Noriega was in full control. Washington, moreover, was not unhappy that Arias—known for his anti-Americanism—had been deprived of his election triumph. The Panamanian Defense Forces (PDF), as the National Guard was now named, was growing in strength. It was taking over all aspects of Panamanian public life, including the immigration department, the civil aeronautics administration, the railroads, the traffic department, and even the passport bureau. Under Noriega's command, the defense forces also received more than $32 million in U.S. military aid, which made it possible for Noriega to expand and modernize the force from about 10,000 to more than 16,000 in just about four years.[16]

It wasn't until the summer of 1987 that widespread protests against General Noriega's rule broke out in Panama. These came in the wake of revelations by a high officer in the PDF of corruption in the armed forces and of the general's role in the exceptionally brutal murder of one of his fiercest critics. In February 1988, Noriega, one of his top officers, and fourteen other people were indicted by a Miami grand jury. The twelve-count indictment for racketeering stated that Noriega had made more than $4.6 million by turning his country into a vast clearinghouse for drugs and for money that was tied to the Colombian cocaine trade.[17]

Suspicions that Noriega was involved in the drug trade went back as far as 1972, and by the late 1970s U.S. officials had no doubt that he was responsible for helping to move large amounts of cocaine from Colombia to the United States.[18] But the Carter administration did not want the case to go forward, especially at a time when Noriega had recently done the administration a favor by giving refuge to the Shah of Iran.[19] Noriega established even closer relations with the U.S. government in the 1980s. He worked for the Central Intelligence Agency, providing the U.S. Drug Enforcement Administration with information on Havana and lower-level drug dealers. But he was a double agent several times over. He gave information and underground help to Fidel Castro. He supplied arms to the Marxist Salvadoran rebels and to the Salvadoran government, to the Sandinistas and to the contras.[20]

By the end of 1987, the Reagan administration seemed finally to have concluded that Noriega was expendable. On February 25, 1988, the president of Panama, encouraged by Washington, tried to depose Noriega. He failed. Over the next twenty-one months, the Reagan and Bush administrations applied economic and political pressure to rid Panama of the drug lord. Finally, in December 1989, after an American officer had been killed by the PDF and another U.S. officer and his wife roughed up, George Bush sent in U.S. troops to install as president Guillermo Endara, who had been legally elected that May but never allowed to hold office.

With Noriega gone, the most serious problem—even

greater than rebuilding the shattered Panamanian economy—
has been to reform the immensely corrupt Panama Defense
Forces, now labeled the Panama Public Force. Over the years,
the armed forces and its individual officers have been reported
to hold interests in about 60 percent of Panama's commercial
enterprises. Military officers collected their own taxes on mer-
chandise in the Colon Free Zone on the Atlantic coast, on savings and loan companies, and even on profits from prison
labor. "You can't make a mafia into a professional army," said
one Panamanian dissident in 1987.[21]

It might be possible, however, to re-create the military as it
was before the 1960s, by training a relatively small force for essentially police functions. The demilitarization of Panama is more
likely to ensure democratic institutions than a sizable army; this is
what democratic Costa Rica has lived with since it dissolved its
armed forces in 1948. As the late President Arnulfo Arias report-
edly said while reviewing a parade of the armed forces at his
inauguration in 1968 (before being deposed ten days later),
"Who are they supposed to fight? The Panamanian people?"

The future of Panama necessarily involves the future of the
U.S. military presence there and the administration of the
Panama Canal. It would seem ill advised on the part of the
United States to move the Southern Command from Panama
until democratic institutions are firmly established in the country. But there is a good case for placing the military command
in the United States. This is what the former head of South-
com, General Paul Gorman, urged when he testified before a
Senate subcommittee. Using facilities in Florida and Texas for
operations in the region would be more secure, and, except in
Panama itself, troops would arrive on the scene just about as
rapidly as needed.

The canal remains a vital international waterway, even in
an era when the United States maintains a two-ocean navy.
While it is too small for supertankers, (in fiscal 1987) the canal
produced some $330 million from tolls and revenues. More-
over, the Japanese, eager to go on using the canal to ship their
growing export trade with Europe and Latin America, have
offered financing—well over $400 million—to widen the fifty-
one-mile-long waterway. It may well be that by the end of the
century, when the canal is scheduled to be fully turned over to

the Panamanians, it will be Japanese board members who will control its running.

I have chosen to review the American involvement in Nicaragua, El Salvador, and Panama because these three countries may well remain, even to the end of the century, inescapable entanglements for any U.S. administration. There are other countries that are also closely tied to the United States by history—notably Haiti, where we not only occupied the island in the first half of the twentieth century and provided aid to the dictatorship of the Duvaliers after World War II, but also during the Reagan years stepped in at a decisive moment to provide Jean-Claude Duvalier with military transport out of Haiti to exile in Europe. Despite the free election in December 1990 that brought Father Jean-Bertrand Aristide to the presidency, durable democratic institutions in Haiti remain problematic, as witness the coup the army mounted against Aristide less than a year later. The tragic history of that nation, after it wrested its independence from the French during the Napoleonic period, gives little grounds for optimism on this score. Moreover, the United States can play only a limited role in encouraging a permanent transition to democracy. We can offer investment if warranted, and we can most certainly refuse to offer material support to a repressive military. But it is neither our duty nor our destiny to impose democracy there.

The greatest danger for the United States is our Wilsonian impulse to set things right in Latin America. America's appalling drug problem has led too easily to the belief that military intervention in South America can prevent the flow of cocaine to the United States by going to the source. In this instance, by utilizing U.S advisers to train the Peruvian army to fight drug-traffickers, we run the risk of becoming involved in fighting the guerrillas, known as the Shining Path, who have fought for years in the Andes. To intervene in a civil war, with unclear objectives, might convert an insurgency against the Peruvian government into a nationalistic movement against U.S. intervention in Latin America.[22] Any victory in the drug war will have to be won at home, though the military may well see their future in fighting "low-intensity" conflicts in the Western Hemisphere; yet it is in just such conflicts that U.S conventional forces have fared so poorly, from Vietnam to El Salvador.

Nonetheless, the Caribbean-Central American economy will always be closely tied to the U.S. market. Not only will the United States react strongly against any foreign military presence there that appears threatening, but it will doubtless continue to retaliate against imports from the region that might undercut domestic U.S. producers. Such an economic policy could not be more self-defeating. If we are interested in promoting stability and democracy in the region, we have an obligation to encourage growth and development of its economy. Direct U.S. aid has generally been a poor remedy for Latin American ills. Too often it has ended up in the Miami bank accounts of the wealthy elites. Greater access to the American market is likely to prove a far more effective way of stimulating development.

To its credit, the Reagan administration recognized this when it launched the Caribbean Basin Initiative in 1983, seeking to grant major trade concessions to several states in the region. But Congress gutted the program by continuing to impose trade barriers. For example, there are restrictions on the import of Costa Rican leather, textiles, and sugar—the very products that are central to the health of the Costa Rican economy and, hence, to Costa Rican democracy. Without increased exports to the United States, it will be impossible for the economies of Mexico, Central America, and the Caribbean to flourish.[23]

In addition, Robert Pastor has suggested "a compact" between the United States and the Caribbean countries, but it should come about only "after the region's leaders demonstrate a commitment to new economic policies aimed at promoting exports, expanding investment, minimizing fiscal and current accounts deficits, and providing incentives for return migration by skilled labor."[24] If there are two lessons Washington should have learned over the past two decades, they are these: economic performance does not depend on how much aid and resources a country receives, but how it uses them; and the most effective policies in the region were those that put greatest emphasis on using the country's own material and human resources. In short, self-reliance.[25]

Debt, however, continues to be a major obstacle preventing

the Latin American economies from recovering and thus adding to the overall health of the global economy. Through their efforts to service their debts, the Latin American countries during the 1980s transformed themselves into exporters of capital. As Pastor has written: "From 1982–88, the net transfer of capital *from* Latin America was $178.7 billion. If one adds to that the State Department's estimate of capital flight for the first four years—$100 billion—then Latin America exported twenty-eight times the Alliance for Progress and sixteen times the Marshall Plan in current dollars."[26] Moreover, Latin America reduced its imports from the United States by roughly 60 percent, which meant the loss of more than 1 million jobs here in the United States.

Debt relief tied to economic restructuring, as in the Mexican case, would likely "permit new investment, exploit unused industrial capacity, and, in so doing, stimulate growth for the benefit of both the United States *and* Latin America."[27] To achieve this, political will, not workable solutions, is what is lacking.

So much of the debate over what the United States should or should not have done in Central America in the 1980s centered around the notion that the United States would lose its "credibility" as a superpower if it could not maintain control there. The credibility argument, as former Assistant Secretary of State for Inter-American Affairs Viron P. Vaky said, easily led to the conclusion "that the mere *existence* of a Marxist regime in Central America damages U.S. "credibility" *whether or not* it is linked to Soviet power."[28] If we cannot get support at home for such an interventionist policy, so goes the argument, our credibility suffers.

So, too, with El Salvador. If the army does not reform itself, the process of political democratization will go nowhere. Yet the Salvadoran army may continue to believe it can ignore Washington's pleas for reform precisely because any U.S. administration must believe U.S. credibility will be gravely affected if a quasi-Marxist state comes into being in El Salvador—even after the end of the Cold War. President Reagan incarnated this fixation when he insisted in 1983 that, should American efforts to combat hemispheric Marxism fail, "our credibility would collapse, and our alliances would crumble." In short, America's credibility in North America was inseparable from our overall strategic safety.

In fact, as military historian Caleb Carr has pointed out, "no such inseparability exists. In the pre-nuclear world, when power was measured in conventional military terms, a base of foreign operations in this hemisphere could threaten our security. The nuclear age seriously weakened the links between the genuine security interests of the United States and those of Latin America. And rapid advances in nuclear technology, most notably that of submarines and cruise missiles, in the last 25 years have severed such ties altogether."[29] For example, should the United States and the Soviet Union have come into conflict in the 1980s, Soviet submarines that cruise freely off the American coasts could have inflicted as much damage on us as any Soviet installations in Cuba or Nicaragua. Our commitment to hemispheric anticommunism seems therefore to have been motivated not primarily, but solely, by our obsession with credibility.

Credibility, of course, is central to the conduct of foreign policy. But in the pursuit of the national interest it must be subordinate to, and clearly differentiated from, the pursuit of genuine American security. Credibility is most justly earned by carrying out a solvent foreign policy, which means defining goals honestly and realistically.

In recent years America's credibility has indeed been in question because of the obtuseness of its foreign policy. If the United States continues its long history of relying on repressive local military forces to maintain stability in Central America and the Caribbean, it will not appear to be planning for a future that will allow countries with fragile democracies to grow stronger and hardy democracies to flourish. If it were to concentrate on negotiating security guarantees for the region when needed, rather than on recklessly arming it, then the United States might find that its credibility is actually strengthened, because countries both friendly and unfriendly would recognize that Washington was at last acting with good sense.

PART V

Solvency Regained

14

A New International Economic Order

A the dawn of the postwar era, at that moment of hope when the United States took the lead in fashioning a brave new world that would somehow avoid the iniquities of the past, the global economy was seen as indivisible. Never again would we be faced with competing blocs and beggar-thy-neighbor policies. Never again would the world be tied to a fixed monetary rate based on the absolute gold standard, that *auri sacra fames*, that sacred base metal that John Maynard Keynes denounced in his *Treatise on Money* calling for a more abundant world. Half a century later, we are nonethless faced with a world that may well evolve into three great trading blocs, accompanied by volatile shifts in the exchange rate, both of which will undermine America's ability to cope with the challenge of maintaining its position in the world economy. Yet as the world's leading economic power, the United States has a signal responsibiity to take the lead in fashioning new institutions that will once again promote free trade and currency stability.

Physically unscathed by the Second World War, the United States in 1945 was the greatest creditor and trading nation in the world. We also devised institutions that we believed would benefit mankind and at the same time perpetuate American dominance in the global economy. What was good for America, so we thought, was good for the world, and what was good for the world was good for America. For at least a generation, this seemed to be true. In the 1930s, on the other hand, the great powers had often sought economic advantages for themselves alone, and only rarely cooperated for the greater good. The decade-long Great Depression was finally "solved" by rearmament and World War II.

The lessons of the interwar period, however, were not lost on the postwar architects of American foreign policy—or on the most prominent member of the British delegation that helped create the postwar economic order, John Maynard Keynes. Meeting at the resort hotel at Bretton Woods, New Hampshire, in July 1944, they were quite clear that the United States, committed to a policy of full employment, would run the world economy and, further, would provide the money to do so. They held that demand was the key to growth and that growth was the key to a benign future. In this multilateral world economy, trade and capital would flow across national boundaries in response to the law of supply and demand. They created a gold-exchange standard that allowed foreign central banks to hold dollars that could be exchanged for gold. From then on, the dollar would be the world currency and provide the liquidity to grease the wheels of world commerce.[1]

The mechanisms needed to stabilize the world economy came to be known as the International Monetary Fund and the World Bank. As designed, the Fund was supposed to satisfy Keynes' main objective: easy credit for countries undergoing a temporary balance-of-payments deficit and pressure on surplus countries to adjust. Countries with chronic deficits would eventually exhaust their right to draw from the Fund. The Bretton Woods system also provided for countries to maintain the value of their currencies at a constant exchange rate relative to other member countries. For example, if the exchange rate of a country was too high, a country would find its international reserves depleted in an effort to maintain that rate. A country could change its rate, of course, but if that change exceeded 10 percent, the change had to be approved by the Fund. When the payments position of a country improved, then it would rebuild its reserves and restore its position in the Fund. In this way, the

Fund's resources would be replenished continually and thus made available to other countries. If a country were so irresponsible as to be in chronic deficit, then it would eventually lose its right to draw from the Fund, and at that point the Fund could place conditions on further access to more money. Although it was designed to fuel the world economy, unfortunately the IMF more and more came to fill the role of a policeman for countries that were financially overstretched: it would lay down conditions that generally brought on severe deflationary policies for the country in question before that country was considered creditworthy enough to apply to the Fund once again.

The World Bank, in turn, was prepared to stimulate the reconstruction of Europe and to provide soft loans to developing countries to improve the infrastructure of those countries. Three years later, the General Agreement on Tariffs and Trade (GATT) came into being as a way of encouraging countries to reduce tariffs in order to bolster world trade. The principle was to give favorable treatment to any country adhering to the GATT; in addition, members of the GATT were to meet every so often to cut tariffs to the lowest practical level.

Though the Bretton Woods system never quite worked with the precision it was intended to, world trade boomed in the postwar era. The United States, the world's greatest creditor nation, helped the world's debtor nations grow. For America's economy also to prosper, the United States needed markets and was prepared to advance the money needed to let other countries buy American goods. The then Assistant Secretary of State Dean Acheson said in 1944 before the Congressional Committee on Postwar Economic Policy and Planning, "You must look to foreign markets." As Acheson saw it, "my contention is that we cannot have full employment and prosperity in the United States without the foreign markets. . . . How do we go about getting [them]? What you have to do at the outset is make credit available." Which is exactly what we did.[2]

The United States pumped dollars into the world economy with a vengeance after 1947 when the Marshall Plan went into effect. For the system of global prosperity to work, the United States had to be willing to run a balance-of-payments deficit if it were not to receive more dollars than it wanted. It also had to be willing to build up the European—and later the Japanese—economies so that these countries could export more and more to the United States, which paradoxically was the only way to ensure markets for U.S. exports to Europe. They needed to earn the dollars by selling to us so that they could buy from us. All this free trade was simply dandy as long as they were junior partners in a U.S.-run, world economy.

While never abandoning the notion that free trade would be good for the world in general and America in particular—whose goods were desired by everyone—Washington decided that a new system was needed to allow the Europeans and Japanese temporarily to set up high tariffs and other barriers to free trade. Only in this way could the Europeans, acting as a bloc, grow strong enough to buy American products. Then, in the course of time, the tariff walls would crumble and worldwide trade be revived. (By the 1960s, this is what generally happened but with far stronger competition from Europe and Japan than most Americans had ever imagined, until by the 1980s many American products were not good enough to compete with their foreign counterparts.)

The Marshall Plan was to supply the money that would allow Europe to build up its economy, and, in so doing, to buy American imports. The dollar shortage was to be overcome by American aid, and to this end the United States in the 1950s was willing to run—and the Europeans and Japanese eager to accept—overall American balance-of-payments deficits, largely as a result of U.S. military expenditures abroad. The need for a strong Europe to resist communist aggression—perceived as coming from the Soviet Union and from internal subversion—meant not only a North Atlantic Treaty Organization but American dollars to finance it. In any case, a growing American economy would, everyone assumed, make good these short-term deficits.

Europe was more than happy to accept American military protection in exchange for the prospect of rapid economic recovery, especially since the United States had most of the world's gold reserves by the end of the war—$22.8 billion worth in 1950—more than enough for the United States to back its expanding currency.[3] Thus, though the European central banks were able to buy gold at $35 a ounce at that time, there seemed little need to do so, and when the European currencies became convertible into dollars in 1958, most Europeans were happy to hold their reserves in dollars rather than cash them in for gold. After all, dollars could be invested at decent interest rates, as gold could not, and, more important, an expanding world economy meant that dollars were the currency needed for global trade and investment.

But slowly the so-called dollar gap closed, and the Bretton Woods system was doomed. The Europeans earned more and more dollars from their exports and from U.S. military expenditures and investments abroad. The cost of providing a Pax Americana, first in Europe and then in the Far East, needed a strong American economy with wise investments in new industries to keep up our competitive edge, and a population willing to sustain its consumption by investing in its own future. None of this finally happened.

During the Kennedy and Johnson administrations, our payments deficits continued, which we justified in the light of the obligations we only too eagerly assumed by extending our military protection over others. We also continued to emphasize the need to reduce tariffs on goods at home and abroad. The active promotion of free trade was designed to keep the world market open for exports, particularly American exports, and so start to improve the falling U.S. trade balance and America's new weakening balance of payments position. In the long run, agricultural exports did help our faltering foreign trade, but they were not enough. In 1971, the United States suffered its first trade deficit in the twentieth century. We have never recovered.[4]

As the 1960s were generally a time of relatively low inflation and high growth, most foreigners were nonetheless disposed to have confidence in America's management of the dollar. Dollars were still needed as international money because an expanding world economy required the enormous liquidity provided by the dollar to finance it; yet the very willingness of foreigners to hold dollars, far in excess of the gold reserves the United States was compelled to hold in exchange, set the stage for global inflation. This was the so-called Triffin paradox—the more the dollar was called upon to met the demands for world money, the weaker it became and the less confidence it inspired among international investors.[5] As dollars continued to flow abroad through investment, imports, and military expenditures, however, we went on playing the role of world policeman, which, in turn, threatened the breakdown of the international monetary system.

In fact, the United States was creating its own credit. At some point or other in the 1960s, there was simply not enough American gold (still pegged at $35 an ounce as it had been since 1934) to redeem all the dollars held by foreigners. But most foreigners continued to hold their overseas dollars because they still believed that the United States could redeem them through its own production potential. Even so, some of these so-called Eurodollars were being cashed in for gold, which resulted in a further depletion of U.S. gold reserves. What we were doing was printing dollars to borrow against the promise of an expanding future, in a world of limitless expectations.[6]

Nonetheless, until the dramatic escalation of the Vietnam War in 1966, it seemed likely that the United States would be able to improve its trade position and its productivity, and to contain inflation, while maintaining low unemployment and high growth. This was surely the best of all Keynesian worlds. Instead, the Johnson administration found that it was not possible, as it had been during the Korean War, for an expanding economy to pay for the cost of combat. Johnson's program was to fight a war without inflicting economic pain, in particular the pain of depriving a consumer society of its wants. It was part of that same promise we had believed in when we preferred to invest abroad in order to provide ourselves with goods produced by cheap labor, rather than invest at home to keep our industrial plant modern. Johnson also refused to finance the war by raising taxes. He did not want to provoke a debate on fighting the war, but, of course, that was the consequence of his policy.

Observing these policies, the Europeans at long last began to lose confidence in America's ability to manage its money. Continuing to fight a war in Southeast Asia while maintaining a powerful military presence in the Western Pacific, North Asia, and Europe—and all this in the face of declining productivity, growing inflation, and the expectation of a higher standard of living—how could the United States also preserve a strong dollar backed by gold and hence a stable exchange rate? The answer was: it could not.

In 1971, President Nixon announced that the United States was no longer willing to convert dollars into gold. We owed so much abroad that we no longer held enough gold to back the dollars we had printed and sent abroad. Yet, foreigners were expected to go on using dollars as world money. In the 1970s, however, their confidence in the management of the American economy waned, and their willingness to hold dollars declined as well. They sought "harder" currencies—the West German mark and the Japanese yen, a recognition of the value of German and Japanese productivity and low inflation. The depreciation of the dollar on the foreign exchanges further fueled domestic U.S. inflation, as the cost in dollars of imported goods rose during the Ford and Carter administrations. Domestic manufacturers, far from taking advantage of the falling dollar to undersell imports, took advantage of higher prices for competing imports to raise their own prices. With billions of dollars still held abroad, foreigners lived in fear that these would someday prove worthless.

By 1978, for the first time in memory, the United States faced a situation where the weakness of its currency on the world's exchanges affected the domestic life of its citizens. We could no longer run the world economy as we chose. At the end of the 1970s, the Federal Reserve, in order to curb inflation and defend the value of the dollar abroad, tightened the money supply. Interest rates at home rose drastically, at the expense of American borrowers and business investment. These measures, and others designed to curb the expansion of domestic credit, temporarily managed to shore up the dollar, but they also produced a severe recession in the early years of the Reagan administration.

The Reagan years were characterized by high real interest rates that were needed to continue to attract foreign capital to finance the deficit Reagan was running up by his program of indiscriminate military expenditure and regressive tax credits. American consumers went on a buying spree and private savings remained low, despite tax cuts. Our debt to foreigners is now likely to exceed half a trillion dollars by 1992. But the dollar, which would decline precipitously if interest rates were to fall to the low levels of the 1960s, remained the world's currency. An overvalued dollar also meant cheaper imports, and this further fueled the consumer's desire to spend, which helped the United States to run a consistent trade deficit of over $100 billion a year, coupled with budget deficits of well over $200 billion a year. High interest rates also meant that the Third World, which had borrowed money at low interest rates in the 1970s, suddenly found itself burdened with enormous servicing charges as interest rates on its debt increased.[7]

Committed ideologically to free trade, the United States sought ways to control the exchange rate so that the dollar would become cheaper and hence our goods less expensive (i.e., more competitive) on the world market. But we could no longer act alone as the locomotive of the world economy. We could no longer manage the world economy by supplying it with dollars that were backed by gold or the equivalent in productive goods. The Reagan administration was therefore forced to seek agreements with the Japanese and the Europeans (mainly the West Germans) on setting the value of the dollar. The aim of the United States by the mid-1980s was to establish stable currency values and trade flows, exactly the same aims the postwar planners had in 1944. But in seeking this, America now needed the cooperation of the other powerful trading nations in order to allow it to shift from what might be called debt-led growth— growth stimulated by the demand created through running massive federal deficits—to export-led growth—which meant offering cheaply priced U.S. products to foreign buyers.[8]

At a meeting at the Plaza Hotel in New York in 1985, the foreign central bankers did agree to support the dollar at given rates of exchange, with the idea of letting the dollar sink to a lower exchange rate so as to avoid any free fall of the dollar. By then we were begging Japan and Germany to become the engines of global demand, which presumably would result in rectifying our trade deficit. While the U.S. trade deficit did decline as a result of the depreciated dollar, its decline was not so marked that the Reagan and Bush administrations were satisfied. In 1990, the trade deficit was still running over $100 billion a year. Neither Japan nor Germany was willing to risk inflationary pressures in order to import enough U.S. goods to change the balance of trade in America's favor; and the Latin American countries were still so burdened with debt that they were forced to pursue export-driven economies and were discouraged by conditions laid down by the IMF from importing.

Confident that foreigners have little choice but to prop up the dollar, Washington has insisted that Germany and Japan protect the dollar. Moreover, the United States believes that if its foreign debt becomes intolerable, it can simply pay back foreign claims on the dollar in devalued currency. This is possible for the moment, largely because of the policy of the oil-producing nations (OPEC) of establishing oil prices in dollars. Since almost every country needs to buy oil, almost every country needs to hold dollars. But should OPEC decide that the risk was too great to hold its asset in a weak currency, it would doubtless shift to holding the yen or the D-mark or the European Currency Unit. This would undoubtedly speed a movement way from other countries' using the dollar as a reserve currency and as world liquidity.

If this occurred, the dollar would plummet, and the U.S. economy with it. Americans would then have to earn foreign exchange to pay their fuel bills, and imports would drop. In addition, our foreign debt would further cut into our living standards. We would be poor debtors in an unforgiving world.[9] At that point, America would find itself unable to control its economy except either by risking high inflation or by using high interest rates to ensure a deflationary economy, which in turn would dramatically reduce our standard of living.[10] Moreover, a return to very high interest rates that characterized the U.S. economy at the beginning of the Reagan administration would cause a further default by Third World countries on repaying their foreign debts. And once again the United States would be accruing an ever greater foreign debt as high interest rates sucked in foreign capital. This is hardly the world the architects of Bretton Woods envisaged—unstable exchange rates accompanying drastic revisions in the value of the dollar in order to rectify growing U.S. trade deficits.

As the century draws to a close, then, the United States is like the proverbial borrower who is in hock up to his eyeballs: if he owes too little, the bank will demand its money back; if he owes a great deal, the bank has a vested interest in keeping him afloat. The power of the debtor, if the debtor is America, is so great that it is always a temptation for Washington to take unilateral action at the expense of the creditor. Over the long term, however, even the creditors will begin to move away from using the dollar as world currency and the United States will find itself at the mercy of the outsider, as it seeks foreign markets and deliberately reduces the standard of living of its citizens to do so.

Given the high levels of fiscal deficit that the United States has been running for years, a severe global recession would probably preclude the now conventional resort to government spending as a stimulus to pulling out of the recession. At bottom, as long as the United States goes on living beyond its means, it will build up a greater and greater debt burden that will have to be repaid in ways its creditors will find equally unpleasant: either by running a trade surplus of something like $50 to $100 billion per year (which the European or the Japanese will be expected to absorb) or by paying back the interest with depreciated dollars.[11]

From this perspective, how will the world be made safe for free trade? The answer is: it probably won't—unless new institutions are created to buttress the old. The aim of Bretton Woods was to ensure global growth. With the United States no longer able to compete as it once did in the global marketplace, with the European Community attempting not only to strengthen itself as a single market but also to incorporate the changing economies of Eastern Europe, and with Japan seeking markets wherever it can find them, the likelihood is the emergence of large regional trading blocs. As the then Prime Minister Margaret Thatcher pointed out at the 1990 economic summit in Houston, "there are three regional groups at this summit, one based on the dollar, one based on the yen, one based on the deutsche mark."[12] The question is, will these blocs be protectionist, in a hostile competition for predominance? Or will they be compatible with an open global marketplace?

As long as U.S. policy is aimed at forcing the Europeans and the Japanese to accept enough U.S. exports to provide us with a trade surplus to pay our debts, the response is likely to be a rebuff.[13] In addition, by insisting that Third World countries maintain export-driven economies, the United States may in effect push these countries to seek trading havens in Europe and Japan. A world of managed trade and bilateral arrangements, the world Bretton Woods was designed to eliminate, may come into being once again.

An integrated and expanded European Community might have little interest in buying more goods from America. Its agricultural needs can easily be fulfilled within the Community, and with the opening of Eastern Europe and the Soviet Union, Europe's traditional policy of tying together imports and exports is likely to be reinforced. The Europeans, moreover, are tending to band together to develop new technologies, and to erect tariff barriers and other impediments to improve and expand their consortia. Already the Europeans have severely restricted imports from Japan, and even when the Japanese have built automotive factories inside the Community, as they have in England, the Community sets up new rules that require more and more components of cars constructed in Europe to use European-made parts. The Japanese cars eventually may make some dent in the European market, but they will be largely European-made, while the profits earned can be repatriated to Japan or elsewhere. Moreover, as the United States draws down its military forces in the European theater, the implicit—and sometimes explicit—leverage Washington exerted over European economic policies will disappear.

In the Pacific, Japan exports capital, high technology, and capital goods in exchange for raw materials and low-end manufactured goods from East Asia. In addition, Japan imports raw materials and food from the United States, Australia, New Zealand, and, to an ever larger degree, South America.

The weakest of the emerging blocs is likely to be the North American-Caribbean grouping. As I have pointed out, a U.S.–Mexican free-trade zone is likely to join with the U.S.–Canadian free trade area. In addition, by opening the North American market to the developing countries of the Western Hemisphere, the United States and Canada will make it possible for them to export those products, such as textiles and agriculture, where they still have a comparative advantage.

While the Bretton Woods system that depended on the actions of the United States to promote growth and harmony in the world economy is no longer viable, neither is the system of trying to coordinate policies among the Europeans, Japanese, and Americans on an informal basis. Domestic politics generally undermines even the best will to international order. What are needed now are new and revised international institutions that will meet the needs of a global economy so that the great regional trading blocs need not engage in competitive economic wars. It will not be enough simply to enhance the existing resources of the Fund. An expanded version of the International Monetary Fund, a kind of supranational central bank, will be required. To deal with a global recession, for example, no international agency is now capable of strategic planning and rapid action, and individual nation-states may not be able to take action on a grand scale. As the writer Walter Russell Mead has pointed out, "Nation-states are now standing in relation to the whole in much the same way as the firm does to the nation-state."[14]

There is already a kind of model of a new international monetary system operating on the regional level among members of the European Community. The European Monetary System (EMS) has been in effect for about ten years and now hopes to produce a European central bank and a European currency by the mid-1990s. The exchange-rate mechanism of the EMS ties its members' currencies together in a rather loose system. Exchange rates are permitted to vary within relatively narrow bands: when a currency appears about to move outside the band, the country concerned is forced to take action—generally a change in interest rates—to prevent it. As West Germany is the strongest member of the EMS and pursues a policy of monetary restraint, interest rates have tended to follow its lead, and this has brought inflation in countries such as France, Italy, and Ireland down toward the low rate of West Germany's.

The problem with adopting such a system on a global level, as *The Economist* has pointed out, is that any exchange rate model that includes the United States will have to take account of America's disproportionate economic and political power, as well as "its un-German willingness to take chances with inflation." Nonetheless, a global currency union (which would allow countries to accumulate international debts and credits within certain limits) based on an independent world central bank might stand a better chance than "a system that relied on the countervailing political weight of other governments committed to low inflation." In *The Economist*'s scheme of things, a world central bank would also have to have a common global currency. A common currency "would deliver the benefits of greater integration of world markets at the cost of ruling out exchange-rate adjustments. But its inflation rate would depend on the system's central bank."

Even if it seems too utopian to imagine the emergence of a single world currency, it is not out of line to imagine the existence of a stable international unit of account. Such a unit might be worth its weight, not in gold, but in a basket of goods. This would mean promising to redeem the unit for however much gold (or some other asset) is needed at current market prices to buy the basket.[15]

Borrowing from John Maynard Keynes, Mead has proposed an international central bank along the lines proposed by Keynes in his 1930 *Treatise on Money:* "Armed with substantial currency and bullion reserves of its own, committed to preserving stable currency values and of maintaining its own currency—the 'bancor'—as an international money of account and settlement, the International Bank would be in a position both to expand credit and to stabilize prices." The world's central banks would agree to maintain the value of their own currencies against the "bancor"; they would have the power to raise or lower the bank's discount rate, to make advances above the quotas available to member banks, and to hold, as deposits, reserves from its member banks. In essence, the International Bank would be able to affect interest rates and credit creation, just as central banks now do for the national economies.[16]

To promote free trade and thus avoid a self-defeating struggle among the three great trading blocs may require yet another new institution. Here again, one might borrow from Keynes, who promoted the creation of the International Trade Organization (ITO) after the war; unfortunately, the ITO came to naught, since the United States preferred to run things as it saw fit. But the GATT, which became America's preferred instrument to deal with trade policy, has concentrated largely on lowering tariffs among states and has never been able to resolve questions on how to harmonize differences among domestic economic policies, or provide rules of the road. Yet in competing for the global market, the varying conditions under which goods are produced and sold in different countries make free trade more an ideal than a reality. Free trade will not come about by ignoring these differences. The European Community, on the other hand, has worked successfully to reduce barriers to trade and expects to accomplish far more along these lines after 1992.

Of course, global trade is far more complex than intra-European trade, but Europe's achievements in this realm demonstrate that the world would benefit from a representative trade body working out the rules of the road on the basis of consensus, if not of unanimity. The goal should be clear: "goods and services produced in accordance with international standards should enjoy most-favored-nation treatment" and "goods not produced in accordance with these standards" should be subject to internationally agreed-upon restrictions or penalties.[17] The analogue for the role of the International Trade Organization in the world economic system—vesting control of international trade in an international body—is the Interstate Commerce Commission here in the United States.

When we talk about tending to the needs of the global environment, putting an end to drug trafficking, ensuring a safe and steady flow of resources, looking for common solutions to health problems such as AIDS and cancer, and accommodating ourselves to the 5 to 6 billion more people expected to be born by 2010—in short, when we consider the magnitude of what has been called the new global agenda—we must keep in mind the imperatives of the global economy. None of these problems can be dealt with successfully in the face of a Great Depression or even violent and volatile swings in the world economies.

A post–Cold War strategy for the United States has to concern itself with revising its military commitments around the globe and avoiding new alliance systems, as for example in the Middle East, which would curb its flexibility and strain its resources. But it also requires expanding and constructing new institutions that will manage a radically different global economy from the one foreseen by the architects of postwar America and Europe. A world of great trading blocs with a parallel free movement of capital aided and abetted by instantaneous electronic funds transfers and ever more advanced information systems does not lead to a more manageable, and hence a safer, world. It may do just the opposite and produce disorder leading to chaos, as trading blocs seek advantages one over the other, which in turn could lead to right-wing authoritarianism.

If regionalism is to prove a benign development rather than a catalyst for a world of vicious strife between the rich and the poor, then new institutions will have to be developed to regulate trade rather than simply rearrange the barriers to it; old institutions will have to be redesigned to encourage the creation and expansion of new markets. Our expectations for the good society may no longer be limitless, but we need not lapse into endless beggar-thy-neighbor policies if we can redefine the nature of our hopes and the fulfillment of our desires.

15

The American Mission

THE last decade of the twentieth century seems to be ushering in a triumphant age of democracy, and so validating America's democratizing mission. In much of South America, military dictatorships were displaced by democratically elected presidents. After years of oppression, Argentina, Brazil, Chile, Paraguay, and Uruguay—all chose the path to democracy. Above all, people's revolutions in Poland, Hungary, East Germany, Czechoslovakia, Rumania, Bulgaria, and even Albania toppled corrupt and tyrannical regimes. Fears in the West that Eastern Europe's freedom from the bonds of oppression could result in atavistic struggles among competing nationalisms—which seemed most justified in the case of Yugoslavia—were countered by hopes that the very ties the new republics sought to Western Europe might prove strong enough to reinforce the democratic impulse, as they had with Portugal and Spain a few years earlier.

In the United States, the commitment to a moral foreign policy is very much in the American grain. Jimmy Carter's emphasis on human rights as the heart and soul of U.S. foreign policy had not been so far off the mark. And Ronald Reagan echoed Carter's sentiments when he said, "The ultimate goal of American foreign policy is not just the prevention of war but the extension of freedom—to see that every nation, every person someday enjoys the blessings of liberty."[1]

Americans are, after all, most comfortable with a foreign policy imbued with moral purpose. Even when the pursuit of justice has led to unintended consequences, even when our ideals have concealed from ourselves as well as from others motivations of a darker and more complex nature, we have preferred a policy that at least rhetorically is based on moral purpose rather than self-interest.

Throughout the Cold War, which was largely fought out in the Third World, the mission of the United States has been that of both a crusader and an exemplar.[2] As a crusader, America tried to roll back communism—militarily in Korea when Harry Truman and Dean Acheson allowed General Douglas MacArthur to cross the thirty-eighth parallel and strike north in an effort to unify the country; rhetorically under Dwight D. Eisenhower and John Foster Dulles, when the Soviet-bloc nations of Eastern Europe were led to believe that the United States would intervene to aid them in escaping from Soviet control; both rhetorically and militarily under Ronald Reagan, as the United States sought to arm anticommunist guerrillas in Asia, Africa, and Latin America. (In Vietnam, our most tragic war, rollback was never even contemplated; instead, a classic policy of containment was tried, but in a region of only marginal strategic importance and in support of an ally whose domestic base was fatally weak.)

Historically, America has generally preferred to fulfill its mission as a crusader for freedom by acting alone, without allies. But whether as simply the champion of freedom in a benighted and sinful world, or as an activist seeking to make the world safe for democracy, America has viewed itself as exceptional, ordained to play a singular role in world affairs.

America's assumption that she has a redemptive mission not only in its own territory but throughout the world was perfectly expressed eleven years before the Declaration of Independence, when John Adams wrote in his diary in 1765, "I always consider the settlement of America with reverence and wonder, as the opening of a grand scene in Providence for the illumination of the ignorant, and the emancipation of the slavish part of mankind all over the earth."³ But this paean to American exceptionalism should be set against the more cautionary words of Alexander Hamilton, who warned us to reject "idle theories which have amused us with promises of an exception from the imperfection, weaknesses and evils incident to society in every shape." He asks us in *The Federalist*: "Is it not time to awaken from the deceitful dream of a golden age and to adopt as a practical maxim for the direction of our political conduct that we, as well as the other inhabitants of the globe, are yet remote from the happy empire of perfect wisdom and perfect virtue?"⁴ But Hamilton was not heeded. Instead, America was deemed an extraordinary nation—"an asylum for mankind," as Thomas Paine wrote in *Common Sense*.

Ours was to be a unique destiny. As Henry Adams describes Thomas Jefferson's idea of the American mission, the third American president "aspired beyond the ambition of nationality, and embraced in his view the whole future of man. That the United States should become a nation like France, England, or Russia . . . was no part of his scheme. He wished to begin a new era. Hoping for a time when the world's ruling interests should cease to be local and should become universal . . . he set himself to the task of governing, with this golden age in view."⁵

The critique of American perfectability was not long in coming; it came not in political discourse but in the writings of the classic American novelists, most notably Nathaniel Hawthorne and Herman Melville. They knew that this vision of an uncharted world where everything is possible was dangerously simplistic—that, on the contrary, everything is impure, even America, and everything is limited, even American possibilities.

Melville, whose masterpiece, *Moby Dick,* was dedicated to Hawthorne, at first appears to be the quintessential, optimistic American man of action. "We are the pioneers of the world," he declares in *White Jacket,* "the advance guard set on through the wilderness of untried things to break a path in the New World, that is ours." He is attracted by the purity of an Eden, glimpsed for the first time in the South Sea islands. But later Melville's tales darken. In his story "Benito Cereno," one hears the tale of an American sea captain, Amasa Delano, who comes upon a drifting Spanish slave ship and, innocently, boards it to help. What he does not realize is that the captain, Benito

Cereno, has been taken captive by the slaves, who have revolted and seized the ship. When Delano himself is threatened by the slaves, he asks, bewildered, "But who would want to kill Amasa Delano?" Unwittingly, he has been drawn into the evils of the Old World. Experience, in the guise of the Spanish sea captain, is akin to corruption; the revolt of the slaves is like a rush from darkness into light. Yet, paradoxically, that revolt threatens the enlightened American's life.[6]

The warning against the search for perfection in an imperfect world never became part of the rhetoric of American foreign policy. Moreover, American foreign policy was singularly successful. It ensured American security from the Atlantic to the Pacific and seemed bent on removing all threats to the vulnerable new Republic. A literature in any way tending to subvert the extraordinary freedom of action we had in pursuing America's exceptional destiny was necessarily disregarded— except when read as tales of adventure and gothic mystery. Throughout the nineteenth century the U.S. government rarely admitted anything less than a moral vision of the world, in which Americans, virtuous and right, sought perfection on a continent whose vast natural resources seemed to promise autarky and, hence, invulnerability.

But these blessings did not lead to an isolationist America. On the contrary, the advice of Washington and Jefferson to avoid entangling alliances allowed the new nation to act unilaterally in pursuing an activist and interventionist role in the Western Hemisphere, even when this risked war with England and Spain. (In the period of American expansion overseas, the ideology of American righteousness was able to be spread far more easily throughout the globe as long as America was unencumbered by powerful allies who might actually question American exceptionalism.)

John Quincy Adams, like Jefferson, was a man who believed that America stood as an exemplar of freedom. Arguably our greatest secretary of state, Adams counseled us as follows: "Wherever the standard of freedom and independence has been or shall be unfurled, there will [America's] heart, her benediction, and her prayers be. But she goes not abroad in search of monsters to destroy. She is the well-wisher to the freedom and independence of all. She is champion and vindicator only of her own. She will commend the general cause by the countenance of her voice, and the benignant sympathy of her example. She well knows that by once enlisting under other banners than her own, were they even the banners of foreign independence . . . [t]he fundamental maxim of her policy would insensibly change from *liberty* to *force*. . . . She might become the dictatress of the world. She would no longer be the ruler of her own spirits."[7] The "Adams Doctrine," as it might be called, is not a prescription for U.S. foreign policy, but it does underscore an American approach to the world. [8]

The expansion of the United States across the continent, first through the efforts of Jefferson and John Quincy Adams, and later at the expense of Mexico, was firmly linked to the notion of the American destiny as a moral absolute. But that same special quality of the American nation and society—the result of enlightened government, an extraordinary geographical position, and a wealth of natural resources—also appeared to make the United States an obvious target for the world's aggressive or subversive forces. Even though England was a liberal, parliamentary state, her imperial ambitions seemed to threaten not only American territorial integrity but also the American democratizing mission itself.

By World War I, the United States had extended the boundaries of its interests to the far Pacific. The acquisition of the Philippines not only prevented those islands from falling into the hands of a foreign power, such as Germany or Japan, but also achieved a missionary goal. When William McKinley himself finally came to demand total acquisition of the Philippines, he claimed he was inspired to do so by divine guidance. In truth, he said, he had not wanted the islands, but one night he knelt down at the White House to beg for guidance from Almighty God. He came to realize that it was his duty to take the archipelago and "to educate the Filipinos, and uplift and Christianize them, and by God's grace do the very best we could by them as our fellow men for whom Christ died."[9] The Filipinos, in fact, had been Christians for hundreds of years.

So cruel was the American occupation of the islands, with perhaps as many 200,000 Filipino lives lost, that our behavior there seemed, if anything, to contradict America's moral foreign policy and its sacred duty. The philosopher William James, saddened and angry, wrote a passionate condemnation of what America had done: "There are worse things than financial troubles in a Nation's career. To puke up its ancient soul, and the only things that gave it eminence among other nations, in five minutes without a wink of squeamishness is worse; and that is what the Republicans would commit us to in the Philippines. Our conduct there has been one protracted lie towards ourselves."[10] Mark Twain echoed James' views. Public opinion finally had its intended effect on our subsequent interventions in the Western Hemisphere—from Mexico to Nicaragua, and most recently, to Panama. Never again was our behavior so iniquitous as it had been in the Philippines.

Even Woodrow Wilson's invasion of Mexico in 1914 was done not in the name of empire, but rather to impose a government sympathetic to the United States, preferably a democratic government. During the First World War, Wilson became the very personification of the crusader for democracy, a man who believed that only by interfering in the affairs of other nations could the United States wage its campaign of self-determination for all peoples. Unable to compromise with his domestic opponents over the issue of American participation in the League of Nations, Wilson remained convinced of the unique mission of the United States. In his last speech, made in 1919 when he was urging ratification of the League by the Senate, he spoke of the American soldiers who had died crusading for a new world of democratic nations: "I wish sane men in public life who are now opposing the settlement for which these men died . . . could feel the moral obligation that rests upon us not to go back on those boys, but to see the thing through, to see it through to the end and make good their redemption of the world. For nothing less depends on this decision, nothing less than the liberation and salvation of the world."[11] The Senate, as we know, refused to ratify the League.

America's encounter with Europe on the battlefront and at the conference table only strengthened the conviction of America's exceptional destiny. The interwar period was not, strictly speaking, one of American isolationism—except for our participation in the security systems of Europe, which were finally doomed to failure without U.S. backing. We used troops in Guatemala, Honduras, Nicaragua, the Dominican Republic, and Haiti. But as European penetration of Central America was no longer seen as a threat to the region, the United States moved to withdraw its soldiers and settled instead for a policy of using local forces to maintain order—*and* American hegemony. At the same time, clothed in moral righteousness, Americans were only too ready to sign international agreements as long as they were couched, like the Kellogg–Briand peace pact to outlaw war, in idealistic terms.

It was not until the election of Franklin Delano Roosevelt that the country found a president who combined the idealistic aspirations of the Founding Fathers to create a republic of virtue and their realistic appraisal of the need to seek temporary alliances to ensure our security. Like Hamilton, Roosevelt counseled against the dangers of exceptionalism: "Perfectionism, no less than isolationism or imperialism or power politics, may obstruct the paths to international peace." Like Hamilton's, his warnings were largely disregarded as the Cold War came to dominate American politics.

Indeed, throughout the years of the Cold War, the notion of America as a crusader, as a force for freedom, seems to have become engraved on the national consciousness. But spreading freedom, or making the world safe for democracy, if it is to be America's peculiar destiny, is likely to be a lonely task. America's allies have not shared its missionary zeal. They see America's obsession with spreading, and often imposing, democracy in the Western Hemisphere (witness our invasions of the Dominican Republic in 1965 and Panama in 1989) as a traditional great-power concern to preserve a sphere of influence. Except toward their former colonies in Africa, the European nations remain generally uninterested in the ideological struggles of the Third World. It is likely, then, that the United States will continue to pursue a largely unilateral course if it continues to strive to fulfill its self-appointed missionary role.

The United States will find in the future, as it has in the past, that any crusade for freedom will sorely test its moral, to say nothing of its physical, resources. For example, the need to treat with dictators in order to further American interests has always been difficult to reconcile with America's democratizing mission. It seems undeniable that in the long term American interests are far more likely to be advanced by supporting democracies than by supporting dictatorships. Our role in trying to overthrow the legally elected government of Salvador Allende in Chile in the early 1970s, our CIA-sponsored coup against the legitimate government of Jacopo Arbenz in Guatemala in the 1950s, our restoration of the Shah in the early 1950s—none of these "successes" ultimately redounded to the credit of the United States; all were discovered after the fact and tarnished our image as an exemplar of freedom and democracy. Nonetheless, the national interest has often required that America aid regimes that are distasteful, or even ultimately dangerous—most notably during the Second World War when we were allied to the Soviet Union.

To suggest that the United States temper its missionary zeal, however, is not to say that America should abandon itself to a foreign policy devoid of moral concern. This was the policy Henry Kissinger and Richard Nixon followed to a great extent—a Bismarckian foreign policy that relied on manipulating the balance of power. They aimed to deal with our adversaries without imposing moral strictures, without self-righteousness, but also without moral demands. Nor did they perceive that liberal democracies were less likely to make war among themselves than authoritarian states. But despite the short-term successes of Nixon and Kisssinger, their foreign policy left no legacy precisely because of the absence of any moral structure. The question inevitably arose: Could the United States build a domestic base for its foreign policy unless such a policy contained a significant moral component? The answer—supplied by Jimmy Carter, Ronald Reagan, and even, reactively, George Bush—was no.

The United States has been most successful in the moral aims of its foreign policy when, short of outright war, it has encouraged the forces of democracy and discouraged the forces of tyranny by its example. It was America as the exemplar of freedom and democracy that caused the nations of Eastern Europe to look to the United States—and the West—rather than the Soviet Union as the model they should emulate. Just as it was the United States as the crusader for the "rollback" of communism that rhetorically betrayed the East Europeans when it stood by as Soviet tanks crushed the 1956 Hungarian revolt.

When we show ourselves unable to balance the pursuit of the national interest with the democratizing mission, American foreign policy is led into crisis. No *American* policy can succeed without such a balance. In the late twentieth century, however, both the democratizing mission and the advancement of the national interest may very well be fulfilled by an unstinting support of international institutions committed to collective security and the preservation of peace. For the first time in forty-five years, the five permanent members of the Security Council—Britain, France, China, America, and Russia—are beginning to play the role they were assigned at the founding of the United Nations: as custodians of international order. As the Cold War came to an end, the Security Council worked to resolve conflicts in Cambodia, Namibia, Central America, the Western Sahara, and the Persian Gulf. The United Nations endorsed the importance of free markets, banned ecologically harmful fishing nets, accepted white South Africans' right to safeguards under black rule, and agreed to negotiate a treaty on stabilizing the world's climate by 1992.[12]

The United States, however, seemed reluctant to fully embrace the role of the Security Council as the central mechanism of world order. America's traditional temptation to unilateral action led us to invade Panama and to resist putting our armed forces in the Persian Gulf or Saudi Arabia under U.N. command. We remained seriously delinquent in our payments for the upkeep of the U.N. organization.

Yet peace is often best served by prevention or preemptive action. Sir Brian Urquhart, a former U.N. undersecretary general, has suggested that small nations, if they believe that they are threatened, should be able to appeal to the Security Council for both reassurance and, if need be, practical measures to guarantee their security. During the Cold War, the Security Council was only able to act "after a crisis had occurrred, to defuse the situation and to limit its consequences." The new political climate could now permit the Council to play a far more convincing role both in preventive action and in providing security. This will require, as Urquhart pointed out, "a transition from the sheriff's posse to the beginnings of a regularly established, and respected, international police force, monitoring the implementation of international decisions and agreements."[13] The Soviet Union seems willing to support this role for the Council, perhaps as a less costly way of influencing events as it scales back its military commitments.

The United States was also founded, as historian Arthur Schlesinger, Jr. points out, "on the proclamation of 'inalienable' human rights, and human rights have ever since had a peculiar resonance in the American mind."[14] The democratizing mission—whether by example or by intervention—has been America's answer to the question of how we would fulfill this mission. In this respect, the creation of the United Nations represented Franklin Roosevelt's own commitment to human rights, the embodiment of the four freedoms he proclaimed in 1941—freedom of speech and of worship, freedom from want, and freedom from fear (or military aggression). These principles have been violated by the great powers, most recently by China in its bloody reaction to the student and worker protest in Beijing in 1989—and they are likely to be violated again; nonethless, the 1980s witnesssed a significant and activist role by the United Nations in exposing and condemning the shameful disregard of human rights throughout the globe. The U.N. Commission on Human Rights submitted reports on torture, mutilation, and other atrocities committed by the Soviet troops during their war in Afghanistan. Agencies of the United Nations now regularly investigate complaints, collect and release information, harass governments, and publicize standards of behavior. Other international instruments, such as the World Court and the Council of Europe, also set standards in international law and democratic norms.

Of course, it is unlikely that we will ever see a world in which human rights will be universally assured. The Hobbesian universe where life is "nasty, brutish and short" and nations are engaged in a perpetual struggle for domination, seems deeply embedded in human nature. Nor is there an effective mechanism for enforcement of human rights. Nonetheless, American foreign policy, without falling prey to the illusionary notion that peace pacts are substitutes for the credible use of force, is likely to be most successful when accompanied by strong moral values. These values can be expressed not only by creating a domestic polity that aspires to John Quincy Adams' model that America act as an exemplar of freedom and democracy, but also by embracing by word *and* deed the international institutions that respond to our deepest values. Walter Lippmann, after the Kennedy administration's misguided attempt to overthrow Fidel Castro at the Bay of Pigs, wrote: "A policy is bound to fail which deliberately violates our pledges and our principles, our treaties and our laws." He reminded us that "the

American conscience is a reality. It will make hesitant and ineffectual, even if it does not prevent, an un-American policy."[15]

As we approach the twenty-first century, we should heed once again the words of Alexander Hamilton, who urged a policy for nations that was not "absolutely selfish," but rather "a policy regulated by their own interest, as far as justice and good faith permit."[16] Justice and good faith are not drawn from dreams of perfectability, but neither are they expressions of cynicism. They are instead the practical goals of a realistic U.S. foreign policy. They are the very essence of America's democratizing mission.

16

The Realist Vision

FRANKLIN Roosevelt did not want to go to Teheran. It was 1943 and F.D.R. was already a tired man. But he needed to meet with Stalin personally and outline to the Soviet leader his grand design. He would have preferred Cairo, only a ten-day round trip away from the United States. But Stalin insisted on Teheran, where his lines of communication and security would be under strict Soviet control. En route to Teheran, Roosevelt stopped off at Cairo to meet with Chiang Kai-shek. The president had few, if any, illusions about Chiang's ability and willingness to fight the Japanese; but he was determined to bolster Chiang's morale because he wanted China to play a central role in the postwar world, in particular to give China a hand in the balance of power in Asia. [1]

At this meeting, his first encounter with Stalin, Roosevelt pressed home a view of a world that would fulfill both the idealistic traditions of the United States and his own notion of creating a realistic and durable peace. F.D.R., like many of his contemporaries who had lived through the First World War and had seen the retreat into isolationism of the interwar period, feared that the United States would once again withdraw from its foreign entanglements if the planning for the postwar era was characterized by an approach to world politics that rested too heavily on a moral and legalistic framework. F.D.R. understood the need for balance. "Let us not forget," he warned, "that the retreat to isolationism a quarter century ago was started not by a direct attack against international cooperation but against the alleged imperfections of the peace."[2]

Roosevelt's idea had been maturing during the course of the war. In a private meeting with Stalin, F.D.R. sketched out his idea of what would become known as the United Nations. He saw it as a three-part organization consisting of a general assembly of thirty-five to forty nations; an executive committee of ten nations, including the Big Four—the United States, the U.S.S.R., Great Britain, and China—and a third group, which F.D.R. called the "four policemen," in which the Big Four "would have the power to deal immediately with any threat to the peace," or any sudden emergency requiring action. Stalin was skeptical, preferring spheres of interests, but by the end of the conference the Soviet leader declared he was willing to accept F.D.R.'s idea of world organization. Roosevelt was also determined not to restore French rule in Indochina and was generally in favor of dismantling both the French and the British empires. For Roosevelt, the issue was how to achieve idealistic goals realistically, and this could only be done by imposing both a Wilsonian peace, whereby the small nations were represented in the U.N. General Assembly, and the great power peace, embodied in the U.N. Security Council.

Though the notion of the "four policemen" would shortly be abandoned as the Cold War gained sway, the concept lived on in the guise of the five permanent members of the U.N. Security Council—America, Russia, Britain, France, and China. Yet the ability of the five to cooperate was severely hobbled, as the Big Four gave way to the bipolar world of the Cold War that maintained the peace through a balance of terror. Significantly, there was no peace settlement in Europe; instead, Europe was sundered in half. By the 1950s, Britain relinquished its pretentions to being a great power, and China was torn by civil war, then isolated until the 1970s, when mainland China regained the seat that had been held by the nationalists who were confined to the island of Taiwan.

Yet the Rooseveltian vision, almost a half century later, may be emerging as the dominant shape of the postwar system. The five permanent members of the Security Council may not be the five policemen, but in the aftermath of the Cold War they nego- tiated settlements of conflicts in Namibia, Afghanistan, and Cambodia. Even more dramatically, they endorsed the creation of a U.S.-led coalition to use force to expel Iraqi troops from occupying Kuwait, and so to ensure the safe flow of oil from the Persian Gulf. Although the new global balance of powers is already reflected to some extent in the five permanent mem- bers, were Japan to be allowed to join them and Britain and France to be replaced by the European Community, this would reflect the realities of the the twenty-first century.

None of this could have happened without reconciliation between the United States and the Soviet Union, exactly what F.D.R. had hoped to achieve. Nothing could be done, Roosevelt believed, if there was conflict among the great powers, and this effectively meant between the United States and the Soviet Union. In his last report to Congress after the Yalta conference, in March 1945, he thought he had reached agreement with the Soviet Union and could now envisage "the beginnings of a per- manent structure of peace."³ For Roosevelt, the United Nations was the means by which the United States would be bound to a continuing involvement in world affairs.

The stable structure of peace that F.D.R. hoped for rested on four pillars: a significantly expanded role for the United States abroad; an accommodation with the Soviet Union; a major role for postwar China; and the end to colonialism. All of these aims have at last been achieved.

Haunted by the arms buildup and the self-defeating eco- nomic conflicts of the 1930s, Roosevelt's internationalism also encompassed a world in which free trade would lead to an ever- expanding prosperity. With the war still under way, a series of conferences, mostly initiated by Washington, began to shape the world we have lived in for the past half-century—Bretton Woods (which, through the establishment of the International Monetary Fund and the World Bank, provided for currency sta- bilization); Dumbarton Oaks (which drew up plans for the United Nations); Hot Springs (for food and agriculture); Wash- ington (for relief and rehabilitation); and Chicago (for civil avi- ation).⁴ In calling for free markets, full employment, and human rights, the Rooseveltian vision became the postwar vision for the Western world.

It was also the failure of the postwar victors to implement the peace that brought about the breakdown of the postwar order F.D.R. had hoped for. In the Declaration on Liberated Europe, which Stalin, Churchill, and Roosevelt signed at Yalta, the Big Three promised that "the liberated peoples" of Eastern Europe would be permitted "to choose the form of government under which they will live," "to create democratic institutions of their own choice," and to establish "through free elections . . . governments responsive to the will of the people." The Yalta accords, in short, called for the democratic Eastern Europe that may be coming into being—after nearly a half century when Europe was split into hostile ideological camps, precisely because Stalin broke the pledges he had given in the Crimea in 1945.

Is the world that promises to emerge after the Cold War necessarily going to be a peaceful one? Probably not. The struggle for power among nations does not end simply because the great powers are able to compose their differences. There is no indication in the last decade of the twentieth century that war as an instrument of policy is obsolete. The Iran–Iraq war lasted through the 1980s, with hundreds of thousands dead, and resolved nothing; Iraq's subsequent invasion of Kuwait is another example of the reckoning that force makes upon diplomacy. Substantial, long-fought guerrilla wars were abundant in 1990: in Cambodia, Peru, El Salvador, Ethiopia, Liberia, and Mozambique. The great difference is that America and the Soviet Union no longer compete for ideological allies. Small wars may not be settled easily, but the two powers are at least likely to be disposed to seek settlements in concert rather than exacerbate the conflicts. What one historian has called "the long peace," we have generally called the Cold War. It was a peace, however, that was the peace of Europe. It was a peace because the nuclear balance of terror prevented the superpowers from going to war against each other. Where the lines were clearly drawn, as they were in Europe, the risk of conflict was minimal, because any small war on the continent was seen by Moscow and Washington as likely to escalate, so that inevitably the superpowers would be drawn in and hence risk Armageddon.

In regions where the lines were not drawn—in Korea, in Vietnam, in Southwest Asia—proxy wars were fought, with considerable casualties: neither the Korean War nor the Vietnam War was a skirmish or "police action." In the Middle East, the Arab-Israeli conflict was exacerbated by superpower rivalry. And the 1962 Cuban missile crisis, let us not forget, did threaten to bring the world to the brink of war. War was surely not obsolete, even in a bipolar world. In a multipolar world, the world of the twenty-first century, does this mean that war is more likely than ever to break out? Generally speaking, when the international system is not controlled by a single power—or as was the case during the Cold War, by two superpowers—war is an inescapable feature of independent states seeking power and security.[5] But we are talking about world war. Stability, as we have known it during the "long peace," meant no war between the superpowers, no war in Europe, and, above all, no nuclear war. With the nuclear revolution in military strategy, war between the superpowers was made impossible. But nuclear weapons did not, and will not, prevent wars among the non-nuclear nations, or between nuclear powers and non-nuclear countries.

And yet there are grounds for hope that the consequences of the peace that has come about with the ending of the Cold War may lead to a more pacific world. The spread of democracy in Europe, in Latin America, even in East Asia might bring about a more peaceful international order than what we have known in the past. Liberal democracies, as Immanuel Kant suggested at the end of the eighteenth century, do not make war on each other. Kant hoped that the rise of democratic states, in which a civil society grew up with a free polity pledged to the rule of law, would lead to "perpetual peace" based on a "pacific union" of republican states that would serve the common good. In fact, for over 150 years there has been no case of war breaking out between democratically constituted states. Despite the fact that the mass public can be mobilized for war, the decison to go to war is far easier for a military aristocracy to make than for a democratically elected leader who must answer to the public at the polls or in the parliament.[6] (In this respect, support for neither the Korean War nor the Vietnam War was ever put to a test by a vote in the U.S. Congress; and in both cases, diminishing public support was instrumental in bringing America's involvement in those wars to an end.)

Even in the absence of a universal world of democratic states, however, there is the real possibility that international institutions, preeminently the U.N. Security Council, will take the lead in settling conflict. The change in Soviet policy, which has led the Soviet Union into strong support for the United Nations, makes it possible to use peacekeeping forces with the participation of both the Soviet Union and the United States. In the instance of Iraq's invasion of Kuwait, it was the Soviet Union that suggested reactivating the United Nation's moribund Military Staff Committee. This was the instrument envisaged by the United Nations Charter for carrying out military operations under the command of the Security Council's permanent members.

One of the dividends of the ending of the Cold War in Eastern Europe would be the use of the U.N. peacekeeping machinery, perhaps under the aegis of the Conference on Security and Cooperation in Europe, to head off conflicts that might arise from the outburst of nationalistic self-assertion there after four decades of Soviet hegemony. Tensions between Budapest and Bucharest over the treatment of Hungarian minorities in Rumania; Bulgaria's expulsion of its Turkish citizens; ethnic strife in Yugoslavia, and perhaps even in the Soviet Union—all of these disputes may require U.N. intervention, so that they can be controlled impartially. But above all, this will require cooperation among the five permanent members of the Security Council in giving more autonomy to the machinery of the United Nations and a willingness to back it up both politically and, if need be, militarily.

For the United States, a new internationalism for the twenty-first century not only obliges us to seek to strengthen international institutions like the United Nations, as was our goal at the end of World War II, but also to take the lead in devising new structures to preserve the peace and increase prosperity. Necessarily, the global New Deal begins in Europe, the focus of our energies in the immediate postwar period. With the withdrawal of the Russians from the East, the United States has to redefine its role on the Continent. This requires us not only to continue our nuclear guarantee to Western Europe within the framework of the Atlantic Alliance, but also to work for a new European security system that will include the Soviet Union. It is in the U.S. national interest not to balkanize Europe by trying to exclude the former Soviet satellites from a stake in the security of Europe, and hence in our own security. Inevitably, the end of the Cold War will mean a diminished American military presence in Europe, which is all the more reason for us to play a central role in devising the new architecture of European security.

In this respect, it is not enough to locate a unified Germany in NATO; we need to integrate the "losers"—the Soviet Union and the former Warsaw Pact Nations—in a new European security system. The lessons from history have shown us that it is far wiser to include the losers in a peace settlement than to exclude them. At the Congress of Vienna in 1815, the winners—Austria, Russia, Prussia, and Britain—brought France, the defeated former aggressor, into the Concert of Europe precisely in order to give France a stake in the stability of Europe. This brought near absolute peace for almost forty years. In the 1919 Versailles Conference, on the other hand, the victors excluded and isolated Germany, and what had been a bloody conflict from 1914 to 1918 turned into the first phase of a new European Thirty Years War, which ended by effectively putting Europe under the sway of the United States and the Soviet Union.

To guarantee the peace of Europe, the United States and the Soviet Union will have to continue to play a role in the European balance of power. It is therefore the task of the United States, as well as of the European Community, to invite the Soviets into the European concert and in so doing ensure the presence of both Americans and Soviets in the development of a new security organization that will bring about a new period of long-term stability.

In Asia, in the absence of any overall political or military organization like the European Community or NATO, the only security lies in maintaining the balance of power there. As in Europe, however, it is vital for the United States to play this central role. America, after all, is the only nation that threatens none of the other three great powers of the region—China, Japan, and the Soviet Union. Moreover, all three powers want the United States in retain a significant military, political, and economic presence in Asia. Like Britain's role in the European balance in the nineteenth century, America's role in the twenty-first century is to play the holder of the balance in Asia, so that no one power grows so powerful as to threaten the others.

At the same time, the United States should work to reduce the level of armaments in the Far East. We should be the leader, not the follower, in proposing naval and military force reductions. Far from encouraging the Japanese to rearm, we should make it possible for them to reduce their military buildup. We should seek to involve the Chinese on a whole range of security issues, including nuclear force reductions, as we move to another phase of negotiations with the Soviet Union along these lines. If it is quixotic to expect a disarmed Asia and Western Pacific, it is the better part of realism to work for a more modest level of military and naval forces in the region. In no way should this diminish the weight or significance of the U.S. presence.

The American role in North America and the Western Hemisphere is more delicate than elsewhere. Our traditional response to hemispheric security has been to assert our predominance. For much of the nineteenth century, we feared, not without reason, British intervention. Later, we saw Germany as an interloper, whose aims were to further German aggression. During the Cold War, we were wary of the Soviet Union's ideological ambitions. Today, looking at the hemisphere, there is no foreign power that threatens U.S. security. The central factors of destabilization are from the economic and social ills that bedevil the Latin American countries. The narcotics trade, the sometimes crushing debt burden that leads to poverty and social upheaval, the wave of illegal immigration into the United States—these are the immediate and palpable threats to the security of the United States of America. The tools that we possess are debt reduction, trade preferences, and economic aid as an inducement to democratic norms. In many respects, the most complex foreign policy issues will be precisely those that arise in our dealing with that part of the world that is inarguably at the forefront of our vital interests.

Other parts of the world also gravely affect U.S. interests. At times, certain areas will be more important to us than at other times: for example, Thailand or southern Africa on the brink of a race war. Thus the Third World beyond the Western Hemisphere and East Asia remains an area of serious American interest, but it becomes a vital interest only insofar as we or our allies remain gravely dependent on foreign natural resources. By this definition, the Persian Gulf region is a vital interest.

But while the United States at times may have to use or threaten force to ensure the continued supply of oil, as it has had to do during the Iraq-Iran war and at the time of the Iraqi invasion of Kuwait, it should play a circumscribed role in the Middle East. In the aftermath of the Gulf war, Washington expects to maintain a much larger military presence in the Gulf region than than it had before Iraq invaded Kuwait. But any political or military alliance in an essentially unstable region risks committing Washington to playing the role of a status quo power. Both Syria and Iran, the two other big regional powers, would doubtless oppose this.

At the war's end, it seems difficult to imagine that the sheikhs and emirs of Kuwait will able to rule as they once did. Saudi Arabia, less a country than a family fiefdom, is not likely to remain untouched by demands for secular democracy. The Bush administration's Wilsonian moralizing will sound increasingly hollow if the United States simply committed itself to the internal stability of Oman, Bahrain, Qatar, and the United Arab Emirates—entities that have shown little regard for democratic norms. The United States should have no place as the guardian of the old order.

A more promising policy for America to adopt would be to use its political, economic, and military power to play a relatively modest role in the region. We can retain a residual naval presence in the Gulf and, of course, a base in the Indian Ocean at Diego Garcia. We also have basing rights in Turkey, which remains a member of NATO. We can work for arms reductions, and keep some quantities of war materiel in Saudi Arabia and Turkey. But Washington would be wise to avoid entangling commitments in a region of marginal strategic significance. The Gulf war need not imply a deeper involvement by the United States in the area; the other big powers—the Soviet Union, Britain, France, Germany, and Japan—all have at least as great a stake as the United States in the stability of the region.

Above all, the United States must become far less dependent on energy imports. We now use more than twice as much energy to produce one dollar of gross national product as France, Germany, or Japan. By achieving energy efficiency levels similar to those in Europe and Japan, we will free ourselves from the need to respond quite so readily to the ups and downs of the world oil markets. In so doing, we can convincingly demonstrate that it is up to the Gulf nations to take the lead in devising a security system that will ensure the continued flow of oil. The United States should only back up—not seek to control—the efforts of those nations most directly affected by the political configuration in the Gulf.

At the same time, we must be wary of basing our policies on old premises. In the post–Cold War world, the Soviet Union may very well become a major oil exporter in its search for hard currency. If so, the West's dependence on Gulf oil may lessen, and this, too, could affect oil prices and oil supply. It will certainly affect the politics and policies of the oil-producing states of the Middle East and is therefore likely to be a new source of instability in an always troubled region.

Instability in the Middle East can also lead to renewed conflict that will affect the security of Israel, a country to which we are tied by historical and moral bonds. But here, too, the end of the Cold War offers new opportunities for peace. The Soviet Union has already made clear to its Arab friends that they can no longer count on automatic Soviet support. In the past, any settlement of the Arab-Israeli conflict has been impeded by U.S.–Soviet rivalry in the region. The Soviets now seem disposed to work with the United States in seeking to resolve Arab-Israeli differences, and while this may not be enough to produce a just or lasting peace between the Arabs and the Israelis, it provides a window of opportunity that was previously lacking should both sides find it in their interest to negotiate.

But any efforts to reshape the Middle East must be regarded with extreme caution. It would be a grave error were Washington to expend too much of its energy in trying to achieve a general reconciliation. In this respect, Washington and Moscow can and should work to discover common ground between Israel and its Arab neighbors, but they should be under no illusion that this will lead to a final settlement. Preventing conflict may be less dramatic than bringing about lasting peace; nonetheless, it would be no small victory for the great powers if this alone were possible. As Middle Eastern

expert Fouad Ajami warned after the coalition's victory in the
Gulf war, "this is not the collapse of the Ottoman Empire."
Instead, "we will find, sooner than we expect, yesteday's eva-
sions and complexities, old friends who can't deliver and old
foes who will patch up their grievances."[7]

Traditional security concerns, in the Middle East and elsewhere, can hardly be separated from the economic issues that
overshadow the emerging global environment. The dangers of
warring trading blocs can easily arise if North America, Europe,
and East Asia fall into protectionist patterns to preserve an illusion of autarky. The need for new and revised international
institutions to preserve the patterns of free trade and ensure
currency stability seems almost self-evident. Yet history teaches
us that an international economic regime is not likely to come
about without the leadership of the preeminent economic
power. Despite close cooperation with the British at the end of
the Second World War, it was the U.S. Treasury and State
departments that prevailed in designing the postwar economic
and financial institutions.

The United States, while no longer enjoying the predominance it possessed in the immediate postwar period, is still the
leading economic power in the world. Adjusted to reflect real
purchasing power, Americans have an average income 7 percent higher than that of their nearest rivals. Japan's GNP in
1989, for example, was still less than half that of America's. As
one of the world's largest exporters, America has a vital interest
in open markets; as for foreign investment, American companies have almost $350 billion invested abroad (measured in
book value; the market value is much higher).[8] For these reasons alone, the United States is the logical country to take the
lead in devising new economic and financial mechanisms, such
as an international central bank and a revamped international
trade organization.

But the hour is late. At the end of the twentieth century,
there are no superpowers. America can no longer impose on
the advanced industrialized nations our blueprint of how to
run the world economy. We need the close cooperation of the
Europeans and the Japanese, or else no new system will come
into being; if that happens, the likelihood is for continued cur-

rency volatility and growing protectionism, even in a global economy where capital flows are allowed to move freely outside national borders. Our ability to run our economy as we choose is already circumscribed by our increasing economic weakness; Japanese productivity grows at three times, and European productivity at twice, the U.S. rate. Our interest rates are affected by whether or not the Japanese and the Europeans will buy U.S government paper: if the rates are too low, foreigners will simply refuse to buy our Treasury notes and look elsewhere. It is axiomatic that America needs to run a solvent economy to have successful foreign policy. And new international economic and financial institutions are the vital building blocks needed to achieve solvency.

But there are few hopeful signs that the United States is prepared to take measures necessary to shore up its weakening economy and rethink its role in the world. The federal deficit grows apace (now swollen by a scheduled bailout for failed savings and loan associations). Yet how to repair the economic well-being of the United States is hardly a mystery: beyond budget reduction, strong measures are needed to raise our educational standards, and increase personal savings and levels of investment in research and development. American businessmen must stop emphasizing short-term profits and concentrate on long-term market shares. Once the military budget is cut, public funds must be used to repair and refurbish our aging infrastructure.

What if these measures are not taken? Well, over time, as Paul Kennedy has reminded us, there is likely to be "just the slow, steady hemorrhaging of American industry, the gradual takeover of U.S. assets by foreign companies, the continued mediocrity of the educational system, poor rates of productivity, an economy whose gnp grew, year after year, but at only half the rate of the economic growth of Japan, a united Germany, China, and other states."[9] Moreover, the growth of interest groups in Washington is causing the emergence of a kind of blocked society, so that it becomes harder and harder to carry out unpopular reforms, as witness, for example, the arguments over the "peace dividend" that was to accompany a reduced defense budget. A grave and palpable economic crisis might well prove the only galvanizing mechanism that would propel

Congress and the president into taking draconian measures that would restore the economy to solvency.

Nonetheless, America has shown an extraordinary ability to seek pragmatic solutions to its most acute problems. F.D.R. came into office in 1932 with very few new ideas on how to cope with the Great Depression. But he was willing to try almost anything, and in so doing set in place the bulk of the programs under which the United States prospered and the wealth of the nation was redistributed in the postwar era.

The collapse of the Soviet empire, the end of the struggle for ideological supremacy between Moscow and Washington, the absolute necessity to rethink our military commitments and force structures, all this presents a signal occasion for embracing a new American internationalism. It would rest on the proposition that a global balance of powers is coming into being. It would require the United States to abandon any pretense to being the only superpower, yet it would preclude a withdrawal into ourselves. It would require us to use all our efforts to sustain the global balance, taking nothing for granted in our ability to remain economically and militarily vigorous. Roosevelt's vision was of a world of democratic nations, united "in a permanent system of general security" and in a freely trading international economy—a vision, if we recognize it, that we can now hope to achieve as a consequence of the peace.

Notes

Introduction

1. For an excellent account of the events in Rumania during December 1989, see "Report from Romania," *The New Yorker*, April 2, 1990. At this writing, the Soviet Union, as we have known it for three-quarters of a century, is no more. It is unclear what will be left of the Soviet empire or what it will be called; therefore, I have continued to use the term "Soviet Union" for the republics that will presumably remain tied to one another in some legal construct.

2. *The New York Times*, December 25, 1989—December 24 dateline.

3. See Stanley Hoffmann, "The Price of War," *The New York Review of Books*, January 17, 1991.

4. See Thomas Hughes, "The Twilight of Internationalism," *Foreign Policy*, Winter 1985–86.

5. See William Pfaff, "Redefining World Power," *Foreign Affairs*, "America and the World 1990/1991," p. 34.

6. The phrase "interest-based" foreign policy comes from Alan Tonelson's article in *The Atlantic* (July 1991), "What Is the National Interest?"

7. See President Bush's address before a joint session of Congress, September 11, 1990—text published in *The New York Times*, September 12, 1990.

8. Tonelson, op. cit., p. 38.

9. All citations are from Tonelson, op. cit.

10. All citations are from Charles Krauthammer, "The Unipolar Moment," *Foreign Affairs*, "America and the World 1990/1991."

11. Cited in *The New Yorker*, January 21, 1991.

12. Cited in Christopher Layne, "Why the Gulf War Was Not in the National Interest," *The Atlantic,* July 1991, p. 68.

13. President Bush's State of the Union message, January 29, 1991, published in *The New York Times,* January 30, 1991.

14. This citation is taken from an unpublished paper, "The Ideological Origins of the New World Order," by Robert W. Tucker and David C. Hendrickson.

15. See Michael Mandelbaum, "The Bush Foreign Policy," *Foreign Affairs,* "America and the World 1990/1991."

16. Ibid., p. 11.

17. Ibid., p. 12.

18. See Pfaff, op. cit.

19. On a central steering group, see Stanley Hoffmann, "A New World and Its Troubles," in Nicholas X. Rizopoulos, ed., *Sea-Changes: American Foreign Policy in a World Transformed.* New York: Council on Foreign Relations Press, 1990.

20. See Michael Howard, "Reassurance and Deterrence: Western Defense in the 1990s," *Foreign Affairs,* Winter 1982/83; and Mandelbaum, op. cit.

21. Pfaff, op. cit., p. 48.

22. Walter Lippmann, *U.S. Foreign Policy: Shield of the Republic.* Boston: Atlantic, Little Brown, 1943.

Chapter 1

1. Robert Gilpin, "American Policy in the Post-Reagan Era," *Daedalus,* Summer 1987.

2. Peter G. Peterson, "The Morning After," *The Atlantic,* October 1987, p. 50.

3. See Gilpin, op. cit., p. 48. Figures for GNP growth from the U.S. Senate Budget Committee, "An Analysis of President Bush's Economic Record," September 19, 1991.

4. *The New York Times,* February 5, 1991.

5. See James Chace, *Solvency: The Price of Survival.* New York: Random House, 1980, Part I. For federal deficit projections, FY 1991, see *The New York Times,* December 7, 1990.

6. See *The New York Times,* February 16, 1991; March 18, 1991.

7. *The New York Times,* February 16, 1991.

8. See Gilpin, op. cit.

9. *The New York Times,* February 5, 1991; August 4, 1991.

10. See Richard Barnet, "Rethinking National Strategy," *The New Yorker,* March 21, 1988.

11. See *Access,* vol. 1, no. 4, September 1987; see also Bennett

Johnson, "Make Allies Share Defense," *The New York Times*, January 8, 1989.

12. See Gov. Bill Clinton, "America Is Buckling and Leaking," *The New York Times*, June 24, 1988. Gov. Clinton suggests a shift in federal policy on tax-exempt bonds to help finance public works investment. See also *The New York Times*, December 14, 1989.

13. See *The New York Times*, June 24, 1988, p. A1.

14. *The New York Times*, September 26, 1991.

15. See the report in *The New York Times*, September 24, 1991.

16. See *The New York Times*, August 16, 1990, p. A24.

17. I am grateful to Caleb Carr for his writings and thinking on the nature of military threats in the 1990s.

18. See Jack Beatty, "A Post–Cold War Budget," *The Atlantic*, February 1990.

19. See Robert W. Tucker, "1989 and All That," in Rizopoulos, ed., op. cit.

Chapter 2

1. See *The New York Times*, April 24, 1990, for this estimate of the U.S.S.R.'s GNP by Victor Belkin, a prominent Soviet economist from the Soviet Academy of Sciences; see also *The Wall Street Journal*, January 23, 1989. U.S. and U.S.S.R. GNP figures for 1989, *World Almanac and Book of Facts*. New York: Tharos Books, 1991

2. See Martin Malia, "The August Revolution," *The New York Review of Books*, September 26, 1991.

3. *The New York Times*, August 24, 1991.

4. "Zhivoe tvorchestvo naroda. Doklad tovarishcha M.S. Gorbacheva," *Pravda*, December 11, 1985, quoted in Ed A. Hewett, *Reforming the Soviet Economy: Equality versus Efficiency*. Washington, D.C.: The Brookings Institution, 1987.

5. *The Wall Street Journal*, February 6, 1989.

6. Source: PlanEcon, *The New York Times*, May 24, 1991; see also *The New York Times*, October 16, 1991.

7. See "Perestroika," a survey in *The Economist* (London), April 28, 1990.

8. Quoted in *The New York Times*, April 27, 1990.

9. See interview with Gorbachev's personal economic adviser, Nikolai Y. Petrekov, *The New York Times*, June 10, 1990.

10. Gorbachev interview in *Time* magazine, June 4, 1990.

11. Ibid.

12. An excellent account of the coup and its consequences can be found in Martin Malia's article in *The New York Review of Books*, cited above.

13. Malia, op. cit., p. 26.

14. Havel quote cited by Malia, op. cit., p. 22.

15. See Robert Legvold's *Gorbachev's Foreign Policy: How Should America Respond?* Headline Series, no. 284. New York: Foreign Policy Association, 1988.

16. Cited by Robert Legvold in his article, "The Revolution of Soviet Foreign Policy," *Foreign Affairs* "America and the World 1988," p. 4.

17. Ibid.

18. Quotation appeared in *Literaturnaya Gazeta*, cited in *The New Republic*, January 21, 1991.

19. The roots of Gorbachev's foreign policy thinking go back to the Brezhnev period; works on "new political thinking" were being published in late 1983 to early 1984. See Allen Lynch, "Does Gorbachev Matter Anymore?" *Foreign Affairs*, Summer 1990.

20. Ibid.

21. *Pravda* and *Izvestia* (Moscow), September 17, 1987.

22. See Toby Gati, "A New Approach to the United Nations: Is It for Real?" (unpublished). See also Toby Trister Gati and Edward C. Luck, "Gorbachev, the United Nations, and US Policy," *The Washington Quarterly*, Autumn, 1988.

23. Gati, op. cit.

24. See *The New York Times*, October 3, 1988.

25. *Pravda* and *Izvestia*, September 17, 1987.

Chapter 3

1. Seweryn Bialer points this out in an unpublished paper on Eastern Europe, January 1990.

2. *The New York Times*, December 4, 1989.

3. Quoted in Bialer's unpublished paper, op. cit.

4. Ibid.

5. *The New York Times*, February 22, 1990.

6. From Vaclav Havel's *The Anatomy of Reticence*, cited in Timothy Garten Ash's chapter, "Does Central Europe Exist?" in his book, *The Uses of Adversity: Essays on the Fate of Central Europe*. New York: Random House, 1989.

7. See Ash, "Does Central Europe Exist?" op. cit..

8. *The New York Times*, July 7, 1991; July 8, 1991.

9. See *The New York Times*, June 24, 1991.

10. The comparison between Havel and Masaryk was made by the former U.S. ambassador to Czechoslosvakia, William Luers, in his

talk before the RIIC Luncheon Seminar Series, "Changes in Eastern Europe: Getting to the Velvet Revolution," March 28, 1990.

11. *The New York Times*, July 7, 1991.

12. The Eagleburger speech was delivered at Georgetown University, September 13, 1989—cited in *The New York Times*, September 16, 1989.

13. See Jack Snyder, "Averting Anarchy in the New Europe," *International Security*, vol. 14, no. 4, p. 31.

14. Cited by James Chace, "Auf Wiedersehen USA," *International Management* (London), June 1990.

15. Cited in "Delors' Superstate," *International Management* (London), March 1990.

16. Figures from Salomon Brothers, 1990.

Chapter 4

1. John Maynard Keynes, *The Economic Consequences of the Peace.* New York: Harcourt, Brace and Howe, 1920.

2. See L. C. B. Seaman, *From Vienna to Versailles.* New York: Harper/Colophon, 1963, p. 128.

3. Wolfgang Berner and William C. Griffith, "West German Policy toward Central and Eastern Europe," in *Central and Eastern Europe: The Opening Curtain?* Boulder, Colo.: Westview Press, 1989, p. 340.

4. Ibid., p. 341.

5. In 1988, nonetheless, West Germany's economic ties were still most dramatically in Western Europe. Fifty percent of West Germany's trade was with the European Community. Eastern Europe, including East Germany and the Soviet Union, accounted for only about 5 percent.

6. Cited by Sidney Blumenthal in "Versailles Diary," *The New Republic*, June 11, 1990.

7. *The New York Times*, February 21, 1990.

8. Cited in *The New Republic*, March 26, 1990.

9. Kohl citation in Leigh Bruce, "Europe's Locomotive," *Foreign Policy*, Spring, 1990, p. 69.

10. Dominique Moisi, the deputy director of the French Institute for International Affairs, put it best: "Nothing is more dangerous than to say to Germans today, 'We fear you.' If we do that, we will create a Germany according to that image, the kind of Germany we would deserve." (*Time*, December 18, 1989.)

11. See Robert Gerald Livingston, "A United Germany's Foreign Policy," Opinion Section, *Los Angeles Times*, July 15, 1990.

12. Ibid.

13. A further discussion of a European Security Organization can be found in Chapter 6.

14. *The New York Times,* July 17, 1990.

15. Livingston, op. cit.

Chapter 5

1. Jean-Jacques Servan-Schreiber, *The American Challenge.* New York: Harper, 1968; Schlesinger citation, see Servan-Schreiber.

2. *The New York Times,* June 8, 1990.

3. *USA Today,* June 28, 1991.

4. Even before the unification of Germany, the European Community produced a world GNP two and one-half times larger than Japan's. In 1987, according to figures from the CIA and the OECD and from *The Wall Street Journal,* February 18, 1989, the European Community's share of world GNP was 22.1 percent, the United States' 25.9 percent, and Japan's 9.4 percent. I am indebted to Gregory Treverton's unpublished manuscript, "America, Germany and the Future of Europe," for a discussion of these statistics.

5. See "Transatlantic Shock," *National Journal,* no. 16, April 21, 1990.

6. *The Economist,* June 24, 1989, p. 18.

7. See Chace, "Auf Wiedersehen USA," op. cit. See also *The New York Times,* December 8, 1990.

8. See "The Global Economy," *Great Decisions 1990.* New York: Foreign Policy Association.

9. See *The New York Times,* October 21, 1990.

10. See Bruce, op. cit.; see also *The Economist,* June 2, 1990, p. 71.

11. See "Delors' Superstate," op. cit.

12. See Mandelbaum, op. cit., Part IV.

13. Ibid., p. 23; see also Howard, op. cit.

Chapter 6

1. *The New York Times,* July 17, 1990.

2. *The New York Times,* June 8, 1990.

3. Problems of CSCE are discussed by François Heisbourg, director of the London-based International Institute of Strategic Studies, in his unpublished paper, "The Common European Home."

4. This can be found in Richard Ullman's book, *Securing Europe.* Princeton: Princeton University Press, 1991, chap. 3. See also *The New York Times,* October 17, 1991.

5. A valuable discussion of the proposed European Security

Organization can be found in Malcolm Chalmers, "Beyond the Alliance System," *World Policy Journal,* Spring 1990.

6. For FY 1991, the fiscal deficit is estimated at $282 billion; for FY 1992, at $348 billion. *The New York Times,* June 16, 1991.

7. See David Calleo's essay, "The American Role in NATO," *Journal of International Affairs,* vol. 42, no. 3, 1989. This estimate is derived from the FY 1985 defense budget. See also Richard Halloran, "Europe Called Main U.S. Arms Cost," *The New York Times,* July 20, 1989; and *The Military Balance 1988–89.* London: International Institute for Strategic Studies, 1988.

8. See David Garnham, "U.S. Disengagement and European Defense Cooperation," a paper prepared for the CATO conference on NATO, April 3-4, 1989.

9. See Daniel Patrick Moynihan, "The Peace Dividend," *The New York Review of Books,* June 28, 1990.

10. See Chalmers, op. cit.; see also John Mueller, "A New Concert of Europe," *Foreign Policy,* Winter 1989–90.

11. See Chalmers, op. cit., p. 235.

12. Michael Howard, "The Relevance of Traditional Strategy," *Foreign Affairs,* January 1973.

13. McGeorge Bundy, "Strategic Deterrence, Thirty Years Later: What Has Changed?" *Survival,* November–December 1979, London: IISS, p. 11.

14. See Hans A. Bethe, Kurt Gottfried, and Robert McNamara, "The Nuclear Threat: A Proposal," *The New York Review of Books,* June 27, 1991.

15. Bundy, op. cit., pp. 593–94.

16. Charles and Clifford Kupchan, "Concerts, Collective Security and the Future of Europe," *International Security,* Summer 1991.

Chapter 7

1. See Selig S. Harrison and Clyde V. Prestowitz, "Pacific Agenda: Defense or Economics?" *Foreign Policy,* Summer 1990.

2. See the speech by Mikhail Gorbachev at Vladivostok, July 28, 1986, as broadcast by Moscow television, FBIS, July 29, 1986.

3. Interview in *Time* magazine, January 3, 1972.

4. When I refer to the Pacific Basin economies, I am deliberately excluding Canada and Latin America. I am including Japan, China, Australia, New Zealand, South Korea, Taiwan, Hong Kong, Singapore, Indonesia, Malaysia, the Philippines, and Thailand. See also Bernard K. Gordon, "The Asian–Pacific Rim: Success at a Price," *Foreign Affairs,* "America and the World 1990/91."

5. See Harrison and Prestowitz, op. cit.

6. See *The Economist*, December 24, 1988.

7. See James W. Morley, ed., *The Pacific Basin: New Challenges for the United States*, Proceedings of the Academy of Political Science, vol. 36, no. 1 (1986), pp. 1, 2, 63.

8. Robert A. Manning, *Asian Policy: The New Soviet Challenges in the Pacific*, A Twentieth Century Fund Paper. New York: Priority Press Publications, 1988.

9. Ibid., p. 6.

10. For the $30 billion figure, see *The Economist*, December 24, 1988, p. 30; Harrison and Prestowitz in their summer 1990 article in *Foreign Policy* put the figure at $42 billion "on direct and indirect costs of [U.S.] Far East military forces."

11. The plan is presented in the Pentagon's publication, *A Strategic Framework for the Asian-Pacific Rim: Looking toward the 21st Century*, U.S. Department of Defense, April 1990. See also William J. Crowe, Jr., and Alan D. Romberg, "Rethinking Security in the Pacific," *Foreign Affairs*, Spring 1991.

12. See Jerry W. Sanders, "America in the Pacific Century," *World Policy Journal*, Winter 1988–89.

13. "Southeast Asia and the Pacific Nations," statement of Vice-President George Bush from undated 1988 campaign literature; cited in Sanders, op. cit., p. 49.

14. See Sanders, ibid., p. 49; and Paul Kreisberg, "Climate Improves for U.S.–Soviet Pacific Talks," letter in *The New York Times*, October 13, 1988, p. 26.

15. Interview with Paul Kreisberg, Carnegie Endowment for International Peace, Washington, D.C., June 26, 1990.

16. See Sanders, op. cit., pp. 54, 71.

17. *The New York Times*, June 27, 1990.

18. Cited in Sanders, op. cit., p. 56.

19. John Greenwald, "Japan: From Superrich to Superpower," *Time*, July 4, 1988, p. 31.

Chapter 8

1. FBIS, July 29, 1986.

2. While no dramatic movement was made to resolve the issue of the northern islands over the next few years (with Gorbachev clearly preoccupied with the momentous events in Eastern Europe), there is every reason to believe that Moscow still wants a settlement with Japan that would lead to greater Japanese investment in Siberia. After the

August 1991 Russian revolution, the settlement is even more likely to come about.

3. FBIS, "Gorbachev on Foreign Policy," Krasnoyarsk, Moscow Domestic Service, September 16, 1988.

4. Address by Mikhail Sergeyevich Gorbachev at a plenary meeting of the forty-third session of the United Nations General Assembly, December 7, 1988. See also *The New York Times*, January 19, 1989; *The Economist*, February 4, 1889; *Far Eastern Economic Review*, March 17, 1989.

5. See Gorbachev's speech, September 16, 1988, at Krasnoyarsk.

6. See this description in *The Economist*, December 26, 1988, p. 37.

7. International Institute for Strategic Studies, *The Military Balance 1980–81* and *1987–88*. London: IISS, 1981 and 1988. About 200 of the Soviet ships are thought to be true combatants, compared with about 160 of the U.S. fleet. Because the U.S. forces, however, are made up primarily of carrier battle groups, military analysts consider them decidedly superior to the Soviet fleet. (IISS, *The Military Balance 1989–90*; Congressional Quarterly's *Editorial Research Reports*, April 20, 1990.)

8. *Time*, March 5, 1990.

9. Crowe and Romberg, op. cit., p. 129.

10. See Alan Romberg, "It's Quiet in Asia, But Not Business as Usual," *The New York Times*, February 16, 1990.

11. Paul Kreisberg, "The United States and China in the 1990s," *Contemporary Southeast Asia* (June 1988), p. 58.

12. *The New York Times*, June 24, 1991; August 18, 1991.

13. *The New York Times*, April 26, 1990.

14. *The New York Times*, September 11, 1991.

15. Sophie Quinn-Judge, "Moscow Looks East," *Far Eastern Economic Review*, June 2, 1988, p. 59.

16. *The New York Times*, April 21, 1991.

Chapter 9

1. See *The New York Times*, March 7, 1989.

2. See Alan D. Romberg, *The Future of U.S. Alliances with Japan and Korea*. New York: Council on Foreign Relations, 1990. Critical Issues series.

3. *Aviation Week and Space Technology*, March 20, 1989, p. 88. For defense budget figures, see also *Strategic Survey 1990–91*, published by Brassey's (London) for the International Institute for Strategic Studies; *Regional Surveys of the World: The Far East and Australia*, by Europa Publications (London), 1991.

4. Text of Assistant Secretary of Defense for International Security Affairs Richard L. Armitage's testimony before the Senate Appropriations Committee on Defense, Washington, D.C., May 26, 1988.

5. Cited in Paul Kreisberg, "Japanese Security, American Policy," unpublished paper.

6. *Strategic Survey 1987–88*, International Institute of Strategic Studies (London), p. 158; *The Wall Street Journal*, July 2, 1990.

7. See James Fallows, "Is Japan the Enemy?" *The New York Review of Books*, May 30, 1991, p. 36.

8. Ibid.

9. Cited by Lawrence Summers, "What to Do When Japan Says No," *The New York Times*, December 3, 1989.

10. Cited by Harrison and Prestowitz, op. cit., p. 62.

11. See Romberg, op. cit., p. 23.

Chapter 10

1. I myself was in Beijing, Shanghai, and Canton from May 15 to June 1 and witnessed the student demonstrations. Many of the points I raise in this chapter reflect my conversations with students and with high Chinese officials whom I spoke with as a guest of the Chinese government.

2. Cited by Nicholas Kristof in "How the Hardliners Won," *The New York Times Magazine*, November 12, 1989.

3. See Paul Kennedy, *The Rise and Fall of the Great Powers*. New York: Random House, 1987, p. 452; see also *The New York Times*, March 27, 1986

4. See A. Doak Barnett, "Ten Years after Mao," *Foreign Affairs*, Fall 1986, p. 37.

5. Ibid.

6. See John K. Fairbank's review of Jonathan Spence's book, *The Search for Modern China* (New York: Norton, 1990), in *The New York Review of Books*, May 31, 1990. I often use the term *Beijing massacre* rather than *Tiananmen Square massacre* because of the evidence showing that most of the killing occurred outside the square, and was more likely to be directed at workers than at students.

7. *The New York Times*, May 13, 1990.

8. Cited in *The New York Times*, June 27, 1989.

9. Andrew Nathan, "Tiananmen and the Cosmos," *The New Republic*, July 29, 1991.

10. See Harlan W. Jencks, "China's Moderates Will Return," *The New York Times*, June 26, 1989.

11. *The New York Times*, May 8, 1991.

12. See Winston Lord, "China and America: Beyond the Big Chill," *Foreign Affairs*, Fall 1989, p. 20.

13. See Hong Nak Kim, "Sino-Japanese Relations in the 1980s," *Korea and World Affairs*, Spring 1988, p. 100.

14. Conversations in Beijing and Shanghai, May 15–25, 1989.

15. See "China's Economy," *The Economist*, June 1, 1991.

16. See Spence, op. cit.

Chapter 11

1. See James Chace, "Inescapable Entanglements," *Foreign Affairs*, Winter 1988/89; see also "Stand Tall: A Survey of South Korea," *The Economist*, May 21, 1988.

2. See Selig S. Harrison, "Prospects for Korean Unification: North Korea in Transition," testimony prepared for a hearing of the Subcommittee on Asian and Pacific Affairs, Committee on Foreign Affairs, House of Representatives, May 24, 1988.

3. *The Economist*, February 10, 1990.

4. *The New York Times*, August 13, 1989.

5. See Romberg, op. cit.

6. See Chace, "Inescapable Entanglements," op. cit.

7. *The New York Times*, October 3, 1991.

8. Cited in "The Philippine Paradox," *Far Eastern Economic Review*, July 12, 1990.

9. Ibid.

10. Ibid.

11. The poll was conducted by Martilla and Kiley, Inc., of Boston, and was cited in the *Far Eastern Economic Review*, March 30, 1989.

Chapter 12

1. Cited in Ernest May, *The American Foreign Policy*. New York: Braziller, 1963, p. 145.

2. See *Manchester Guardian*, August 28, 1983.

3. Assistant Secretary of State for Inter-American Affairs Edward G. Miller, "Nonintervention and Collective Responsibility in the Americas," *Department of State Bulletin* 23, May 15, 1950, pp. 768–70.

4. Jimmy Carter's belief that the military could be the agent of social change recalled the views of Lyndon Johnson and Richard Nixon, and were echoed by Ronald Reagan and George Bush. The real reason for U.S. reliance on elites was the need for internal stability

and the exclusion of foreign influence. See James Chace, *Endless War—How We Got Involved in Central America—and What Can Be Done*. New York: Vintage Books, 1984, Part I.

5. See James Chace and Caleb Carr, "Securing the Hemisphere," in *America Invulnerable: The Quest for Absolute Security from 1812 to Star Wars*. New York: Summit Books, 1988.

6. At the time of the 1980 census, the total number of migrants crossing Mexico's 2,000-mile border into the United States was set at 2.5 million. Projections prepared for the Bilateral Commission on the Future of U.S.-Mexican Relations suggest that by the turn of the century this number will have increased to 5.6 million. In 1986, the number of illegal immigrants seized in the United States grew to 1.8 million, a 50 percent increase over the previous year. Moreover, undocumented workers in the United States sent roughly $2 billion to their families each year, making migration a vital source of foreign exchange. See Lt.-Col. Michael J. Dziedzic, "Mexico: Converging Challenges," *Adelphi Paper 212* (London), Autumn 1989; and William D. Rogers, "Approaching Mexico," *Foreign Policy*, Fall 1988.

7. Interview with Jorge Casteñeda, *Time* magazine, August 7, 1989.

8. The ratio of public external debt to GNP fluctuated between 14.5 percent in 1950 and 9.5 percent in 1960, and up again to 13.9 percent in 1975. See Luis F. Rubio and Francisco Gil-Diaz, *A Mexican Response*, A Twentieth Century Fund Paper. New York: Priority Press Publications, Table 1, p. 3.

9. Dziedzic, op. cit., p. 11.

10. The Economist Intelligence Unit, *Country Profile: Mexico 1988–89*. London: EIU, 1988, p. 14.

11. Rubio and Gil-Diaz, op. cit., p. 18.

12. George Philip, *Mexico's Internal Conflicts: The Risks of a Declining Economy*. London: Institute for the Study of Conflict, 1988, p. 4; as cited in Dziedzic, op. cit., p. 12.

13. Rubio and Gil-Diaz, op. cit., p. 12. The price of a barrel of oil went from $2.08 in early 1973 to $10.40 in December 1974. The second "shock" of 1979–80 pushed the price up to $32 per barrel.

14. Maurice Ernst and Jimmy V. Wheeler, *The Impact of Industrial Country Protectionism on Selected LDCs*. Indianapolis: The Hudson Institute, January 1987, p. 99.

15. Ibid.

16. See Dziedric, op. cit., p. 14.

17. See ibid., p. 10; The Economist Intelligence Unit, op. cit., p. 11.

18. Luis Tellez, chief of staff of the Ministry of Programming and Budget, quoted in M. Delal Baer, ed., *Mexico and the United States: Lead-*

ership Transitions and the Unfinished Agenda, Significant Issues Series, Vol. X, no. 5. Washington, D.C.: The Center for Strategic and International Studies, 1988, p. 8.

19. While Mexico accepted the "menu" approach defined by its creditors, this strategy did not significantly reduce the country's debt service ratios, which at the end of the 1980s remained roughly the same as they were at the outset of the crisis.

20. See Robert Pastor, "Salinas Takes a Gamble," *The New Republic,* September 10 and 17, 1990; see also "The New Model Debtor," *The Economist,* October 6, 1990.

21. In 1988, 64 percent of Mexican imports came from the United States, while 67 percent of Mexican exports go to the United States. See Gavriel Szekely and Donald Wyman, "Japan's Ascendence in U.S. Economic Relations with Mexico," *SAIS Review,* October 4, 1989; CSIS Congressional Study Group on Mexico, *The Congress and Mexico: Bordering on Change.* Washington, D.C., 1989, p. 11.

22. See Bruce Babbitt, "Reviving Mexico," *World Monitor,* March 1989.

23. See Andrew Reding, "Mexico under Salinas: A Facade of Reform," *World Policy Journal,* Fall 1989.

24. See Jorge Casteñeda and Robert Pastor, *Limits to Friendship: The United States and Mexico.* New York: Knopf, 1988, p. 226.

25. Ibid., p. 228.

26. Cited in Pastor, op. cit.

27. Robert Pastor suggests the following argument: "Mexico, for example, embarked on an expensive steel project in the mid-1970s without consulting the United States about possible agreement on access to its market. If such an accord had been negotiated, the United States would not have been able to limit Mexico's steel exports, as it did in 1984." (See Casteñeda and Pastor, op. cit., p. 212.)

28. Susan Kaufman Purcell, "A New Latin America," *International Strategies* (New York), vol. IV, no. 1, Jan./Feb. 1991.

29. Casteñeda and Pastor, op. cit., p. 212.

30. Ibid., p. 216.

Chapter 13

1. "El Salvador needs peace, and the only path to peace is at the negotiating table," Assistant Secretary of State Bernard Aronson told the House Subcommittee on Western Hemisphere Affairs in January 1990. "[L]et both sides commit to come to the bargaining table . . . and to say and negotiate in good faith until the war is over."

2. *The New York Times*, December 27, 1989.

3. *The New York Times*, June 15 and 26, 1989. See also Peter Passell, "For Sandinistas, Newest Enemy Is Hard Times," *The New York Times*, July 6, 1989; and "Nicaragua in Extremis," *The New York Times*, July 12, 1989.

4. See James Chace, "The End of the Affair?" *The New York Review of Books*, October 8, 1987. See also Forrest D. Colburn's book, *Post-Revolutionary Nicaragua*. Berkeley: University of California Press, 1986, for a telling account of how the Sandinistas' doctrines led to economic chaos and decline well before the effects of the contra war were felt. See also James Chace, "Dithering in Nicaragua," *The New York Review of Books*, August 17, 1989.

6. *The New York Times*, October 19, 1990.

7. *The New York Times*, February 9, 1990.

8. A. J. Bacevich, James D. Hallums, Richard H. White, and Thomas F. Young, *American Military Policy in Small Wars: The Case of El Salvador*. Cambridge, Mass.: Institute for Foreign Policy Analysis, Pergamon-Brassey's, 1988, p. 46.

9. Ibid., p. 24.

10. See Chace, "Inescapable Entanglements," op. cit.

11. See William M. LeoGrande, "After the Battle of San Salvador," *World Policy Journal*, Spring 1990; see also *The New York Times*, April 31, 1991; see also James Le Moyne, "Hope Again in El Salvador," *The New York Times*, September 27, 1991.

12. This suggestion comes from "Bankrolling Failure: U.S. Policy in El Salvador and the Urgent Need for Reform," a report to the Arms Control and Foreign Policy Caucus by Senator Mark O. Hatfield (R-Ore.), Representative Jim Leach (R-Iowa), and Representative George Miller (D-Cal.), November 1987.

13. Cited in David McCullough's book, *The Path between the Seas: The Creation of the Panama Canal*. New York: Touchstone/Simon & Schuster, 1977, p. 383.

14. See James Chace, "Getting to Sack the General," *The New York Review of Books*, April 28, 1988.

15. The National Guard stood at somewhat more than 11,000 men when Noriega made it over into the Panama Defense Force in 1983. By the time of the U.S. invasion of Panama on December 20, 1989, the PDF numbered 3,500 tactical army personnel, 2,000 police, 750 traffic police, 480 sailors, and 400 airmen, plus another 9,000 support personnel. Another 7,000 armed civilians made up the paramilitary Dignity Battalions. (*The Washington Post*, December 20, 1989.)

16. Citation from an unpublished report on Panama by the Senate Staff Delegation, December 8, 1987, p. 8.

17. *Miami Herald*, February 5 and 6, 1988.

18. See report of Senator Birch Bayh, chairman of the Select Committee on Intelligence, February 21, 1978.

19. *The Washington Post*, March 20, 1988.

20. See testimony before the Senate Subcommittee on Terrorism, Narcotics and International Communications, February 8 and 11, 1988.

21. *Los Angeles Times*, September 21, 1987.

22. See Tom Wicker, "This Is Where I Came In," *The New York Times*, April 23, 1990.

23. Robert Pastor points out, in his unpublished paper, "Forging a Hemispheric Bargain: The Bush Opportunity," that "the U.S. precipitated a sugar shock in the 1980s worse than the oil shock of the previous decade. By 1987, the U.S. imported only 10% of the sugar that it bought in 1981."

24. Some of these suggestions appear in Pastor's "Hemispheric Bargain." Other proposals can be found in Richard Fletcher and Robert Pastor, "The Caribbean in the 21st Century," *Foreign Affairs*, Summer 1991.

25. Pastor, "Forging a Hemispheric Bargain," op. cit., p. 101.

26. Ibid.

27. Ibid.

28. Viron P. Vaky, "Reagan's Central American Policy: An Isthmus Restored," in Robert S. Leiken, ed., *Central America: Anatomy of a Conflict.* Elmsford, N.Y.: Pergamon Press–Carnegie Endowment for International Peace, 1984.

29. Caleb Carr, "Security Precedes Credibility," *The New York Times*, December 26, 1987.

Chapter 14

1. See Fred L. Block, "Bretton Woods and the British Loan," *The Origins of International Economic Disorder.* Berkeley: University of California, 1977.

2. See Dean Acheson's testimony at the hearings on *Postwar Economic Policy and Planning, Hearings Before the Special Subcommittee on Postwar Economic Policy and Planning,* House of Representatives, 78th Congress, 2nd Session (1944).

3. See David P. Calleo and Benjamin M. Rowland, *America and the World Political Economy.* Bloomington: Indiana University Press, 1973 (paper), pp. 44–45; see also David P. Calleo, *The Atlantic Fantasy.* Baltimore: The Johns Hopkins University Press, 1970 (paper), pp. 84–85.

4. While in 1971 U.S. imports exceeded exports by $2 billion a year—America's first trade deficit since 1893—by 1989 U.S. imports exceeded exports by about $2.5 billion a *week.*

5. The "Triffin paradox" was named after economist Robert Triffin, who proposed as a substitute for dollars a system of international credit, created by an international organization. This idea became embodied in the Special Drawing Rights of the International Monetary Fund. See Robert Triffin, *The World Money Maze.* New Haven: Yale University Press, 1966; see also Calleo, *The Atlantic Fantasy,* op. cit., p. 86.

6. Calleo, *The Atlantic Fantasy,* op. cit., p. 84.

7. See Chace, op. cit.

8. Walter Russell Mead, "The United States and the World Economy: From Bretton Woods to the Bush Team," *World Policy Journal,* Summer 1889, p. 25.

9. See Walter Russell Mead, "On the Road to Ruin," *Harper's,* March 1990.

10. U.S. dependence on attracting foreign money to fuel its economy was amply dramatized in early 1990 when surging interest rates in Japan and West Germany forced the United States to raise its interest rates at the very moment when a weak U.S. economy was showing signs of a further slowdown that could lead to a recession. (*The New York Times,* January 25, 1990.)

11. See Mead, "The United States and the World Economy," p. 39.

12. *The New York Times,* July 12, 1990.

13. In 1989, the U.S. ran a small trade surplus with the European Community (+$49.1 billion). Source: *World Almanac and Book of Facts 1991,* p. 161.

14. Mead, "The United States and the World Economy," p. 429.

15. See *The Economist,* January 6–12, 1990, pp. 21–24.

16. Mead, "The United States and the World Economy," pp. 432–36.

17. Ibid., p. 441.

Chapter 15

1. *The New York Times,* October 7, 1986. Reagan's statement, however, was used to justify an anticommunist crusade that sought to do more than contain the Soviet Union—in Reagan's words, "the focus of evil in the modern world"—it sought to roll back the Soviet threat, in both its ideological and geopolitical expressions.

2. See Robert W. Tucker, "Examplar or Crusader? Reflections on America's Role," *The National Interest,* no. 5, Fall 1986, pp. 64–75.

3. Cited by Denis Donoghue, "The True Sentiments of America," in Leslie Berlowitz, Denis Donoghue, and Louis Menand, eds., *America in Theory.* New York: Oxford, 1988.

4. *Federalist*, no. 6. New York: New American Library, 1961, p. 59.

5. Ibid., p. 101.

6. See James Chace, "How Moral Can We Get?" *The New York Times Magazine*, May 27, 1977, pp. 38, 40; see also Harry Levin, *The Power of Blackness*. New York: Random House, 1958.

7. ·Cited in Walter LaFeber, *John Quincy Adams and the American Continental Empire*. Chicago: Quadrangle Books, 1965, p. 45.

8. I am indebted to Philip Geyelin for this term, in his essay "The Adams Doctrine and the Dream of Disengagement," *Estrangement: America and the World*. New York: Oxford University Press, 1985.

9. The quotations from Beveridge and McKinley are from Stanley Karnow, *In Our Image: America's Empire in the Philippines*. New York: Random House, 1989, pp. 109, 128.

10. William James to Henry Lee Higginson, September 18, 1900—cited in Bliss Perry, *Life and Letters of Henry Lee Higginson*. Boston: Atlantic Monthly Press, 1921, 2:429.

11. See James Chace, "American Jingoism," *Harper's*, May 1976, pp. 37–44.

12. See *The New York Times*, August 5, 1990, E3.

13. See Brian Urquhart, "Beyond the 'Sheriff's Posse,'" *Survival*, May/June 1990.

14. Arthur Schlesinger, Jr., "Human Rights and the American Tradition," *The Cycles of American History*. Boston: Houghton Mifflin Co., 1986.

15. Walter Lippmann, "Today and Tomorrow," *New York Herald Tribune*, May 9, 1961; see also Arthur Schlesinger, Jr., "National Interests and Moral Absolutes," *The Cycles of American History*, pp. 69–86.

16. Alexander Hamilton, *Pacificus*, no. 4, July 10, 1793.

Chapter 16

1. See Robert Dallek, *Frankin D. Roosevelt and American Foreign Policy, 1932–1943*. New York: Oxford University Press, 1979, p. 429.

2. See James Chace, "How Moral Can We Get?"

3. Cited in Dallek, op. cit., p. 520.

4. See Arthur Schlesinger, Jr., "FDR Vindicated," *The Wall Street Journal*, June 21, 1990.

5. See Carl Kaysen, "Is War Obsolete?" *International Security*, 14:4; see also Kenneth Waltz, "The Origins of War in Neo-Realist Theory. *Journal of Interdisciplinary History*, vol. 17, no. 3, Spring 1988, pp. 615–28.

6. See Michael Doyle, "Kant, Liberal Legacies, and Foreign Affairs," parts 1 and 2, in *Philosophy and Public Affairs*, vol. 2, no. 3 and no. 4 (Summer, Fall 1983), pp. 205–35 and 325–53; see also Tony Smith, "Wilsonianism Resurgent? U.S. Foreign Policy in a Democratic Era," in Rizopoulos, ed., op. cit.

7. *The New York Times*, February 10, 1991.

8. See *The Economist*, February 24, 1990.

9. Paul Kennedy, "Fin-de-Siècle America," *The New York Review of Books*, June 28, 1990.

Index